ROBERT O. CANNON, MD MPH

A NATURALIST'S SEASHORE GUIDE

A NATURALIST'S
SEASHORE GUIDE

—common marine life of the northern california coast and adjacent shores—

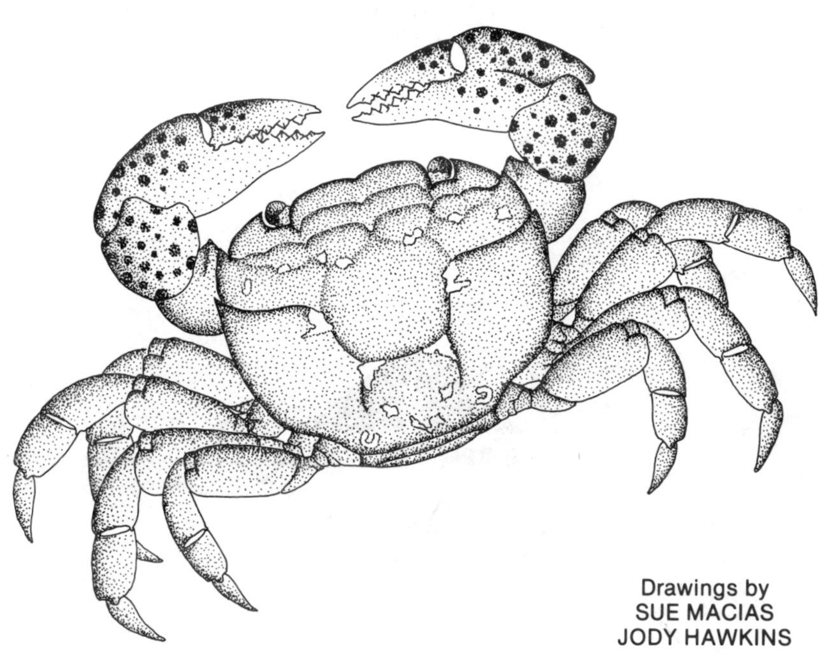

Drawings by
SUE MACIAS
JODY HAWKINS

GARY J. BRUSCA AND RICHARD C. BRUSCA

Gary J. Brusca, Professor of Zoology at Humboldt State University, Arcata, California

Richard C. Brusca, Assistant Professor of Biology, and Curator of Crustacea at the Allan Hancock Foundation, Department of Biology, University of Southern California, Los Angeles, California

Published by Mad River Press, Inc.
Route 2, Box 151-B
Eureka, California 95501

Printed by Eureka Printing Company, Inc.
Eureka, California 95501

ISBN 0-916-422-12-7

Dedicated to Joey, Tony, Alec, Suzy, Carlene, Jamie,
and to natural historians

PREFACE

It is not unusual for brothers to find themselves pursuing similar professions, even though the paths taken by each may not be parallel in time and space. Although separated by age and very different childhood interests, we have not only both become zoologists, but both come to specialize in similar animal groups (amphipods and isopods). For nearly twenty years our paths rarely crossed, save during a few family crises and odd holidays, yet we continued to work, independently, towards similar goals. The credit (or blame as the case may be) for sparking our initial interest in sea creatures must go to Professor David H. Montgomery, who had faith in each of us at different times and offered more encouragement than either of us deserved.

We have seen remarkable changes in the field of marine biology during the past two decades. Both of us have been fortunate enough to study under, and work with, some of the very fine west coast invertebrate zoologists who are true naturalists. The teachings and attitudes of these individuals have instilled in us a respect for all types of life forms. While there is certainly great value in the study of molecules and cell chemistry and the construction of mathematical models that attempt to explain how animals act, we have both been stubbornly persistent in believing that knowledge of natural history is critical to an understanding of what the living world is all about. Our feelings are by no means universal, or even popular in some quarters, but there are answers which can only be found by looking at whole animals, with a hand lens and field notes as tools. In addition to the direct scientific value of this approach, there is something quite special about early morning low tides along the California coast, as the sun rises above the cliffs and scatters the ground fog. It is during such moments that a naturalist is in his closest communion with nature. We envision previous tidepool naturalists such as E. F. Ricketts, S. F. Light, and others out in the field on similar early mornings. And we recall those who took us to the field as students: D. H. Montgomery, in dungarees and flannel shirt, standing elbow deep in a cold pool, bent to a rock with his hand lens; and J. W. Hedgpeth, in peacoat and beret, pipe smoke wreathing his face, perched on a rock humming some odd Welsh tune. There were, and still are, many others.

But the days of the California tidepoolers are numbered. Not only is this sort of science somewhat less popular than it once was, but the coastline itself is slowly being destroyed in the name of "progress" and development. And attitudes have changed. Marine laboratories once inhabited by naturalists tucked away in dusty corners amidst vials and old manuscripts are gradually becoming lost to sterile labs and computers, where white-coated principal investigators send assistants out to collect their experimental animals, or order them from supply houses. We hope that this little book helps to transmit and preserve a bit of the flavor of the natural historians and what they are all about.

The authors gratefully acknowledge the editors of Mad River Press, especially Ms. Virginia Waters and Dr. James Waters. Their valuable and constructive comments on the manuscript have produced a much clearer and more error-free book than might otherwise have been possible. Ms. Sue Macias produced the figures of the invertebrates (other than the few scratchy diagrams by the authors) and Ms. Jody Hawkins drew the fishes, mammals, and algae. We are proud to have their excellent work included in this book. Thanks are also extended to Dr. Robert Cimberg for reviewing the section on intertidal ecology; to Dr. Milton Boyd for allowing us to pester him from time to time for information, for reviewing the ecology chapter, and for supplying some of the photographs; and to Dr. Timothy Lawlor for reviewing the section on marine mammals. Finally we thank Anna Mary and Julie for their patience during the many evenings when our attention was buried in notes, and for their understanding as we often clanked around our respective houses in preparation for a 5 AM low tide.

TABLE OF CONTENTS

CHAPTER I

INTRODUCTION

There have been several field guides and handbooks published during the past ten years or so dealing with the seashore life of the Pacific coast of North America. Most of these works deal with the area from San Francisco south, or with the far Pacific northwest, not specifically with the extreme northern coast of California and southern Oregon. This volume was begun in hopes of filling this need, as well as a personal need on the part of the authors. We assume no background in the subject on the part of the user (although previously gained knowledge will certainly serve to enhance the pleasure of your seashore experience). A listing of selected titles which will provide additional information for the interested reader is presented at the end of this book. Most of the animals and plants chosen for inclusion here are generally common between Fort Bragg and the Oregon border (fig. 1), and it is along that portion of our coast that this work will be most useful. Most of the geographic locations mentioned are on the Humboldt County coast. It should be understood, however, that organisms do not select state boundaries or county lines as their range limits. Many common creatures from as far north as British Columbia, and as far south as Point Conception, can be identified with the information given here, but be aware that the further one strays from northern California, the less appropriate the examples will become. Keep in mind, also, that although the organisms described here are regional, the principles of seashore life are not and should lead to a general understanding of coastal communities nearly anywhere in the world.

We suppose that there are people who find no fascination in the sea, but we do not happen to know any. The impressions made upon those who visit this special part of our natural world have spawned all manner of emotions. Yet, while the seashore may inspire poetry, or serve as a romantic setting for one purpose or another, it also provides homes for myriad legions of animals and plants which are the deserving objects of our lasting interest and careful observation. The purpose of this book is to explain a little about the seashore as a habitat and to discuss the creatures which live there, using northern California examples. Only a small percentage of the total numbers of kinds (**species**) are treated. They have been chosen because of their commonness, ease of identification, or general importance to the community of life on the shore. We have drawn freely from the literature both for examples of organisms and for information about them.

The reader will immediately notice a bias towards animals, especially the **invertebrates** (those without backbones). This situation is in part a result of our personal backgrounds and prejudices. Also, the

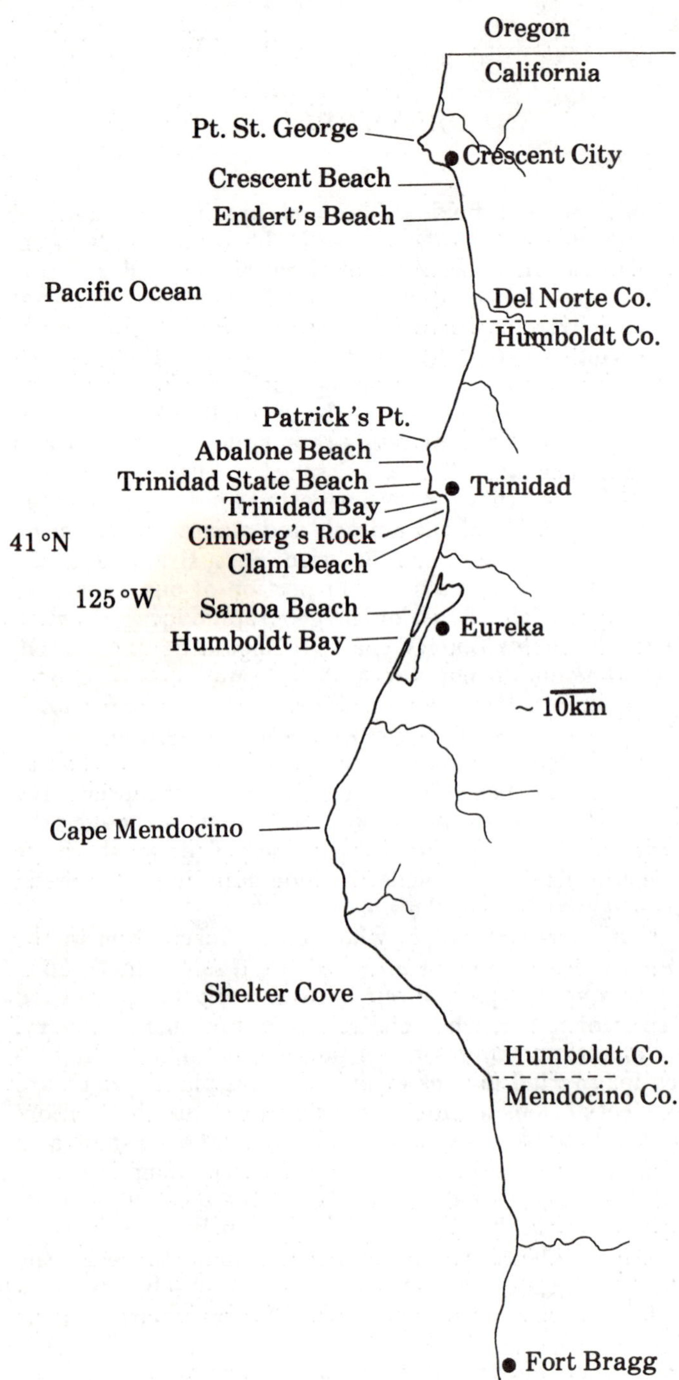

1. The northern California coast

dedicated beachcomber will find that the majority of the animals seen along the shore are indeed invertebrates. Because of these restrictions you will find many organisms which are not included here and cannot be specifically identified with this handbook. In addition, the careful observer will occasionally come upon situations or conditions which contradict the generalizations presented here. In fact, for the dedicated aficionado of seashore life it is a rare trip to the beach that does not uncover a relationship or creature not previously encountered. Treat these discoveries and exceptions to the rules not as frustrations, but as confirmation of the infinite complexity and diversity of life, and as reasons for further study. It is this element of surprise that draws us back, time and again, to the California coastline. Always remember that while the seashore is the most accessible part of the ocean to people, it is an area that is still far from being completely understood.

While it may be argued that the "ecology craze" has gotten out of hand in some respects, there is no exaggerating the pressure on our shores over the past decade. By virture of its accessibility, the seashore is subjected to frequent visitation by ever-increasing numbers of people, for all sorts of reasons. Unfortunately, many of these visits involve activities which are harmful to the coastal environment in one way or another (casual beachcombers who inadvertently litter, developers whose corporate greed would cover the shore with concrete for hotels and marinas, dune-buggy enthusiasts who tear up delicate habitats in the name of sport, bus loads of students who overcollect under the direction of well-meaning but uninformed professors); the list of offenders goes on and on, some are obvious and some are not.

Assuming that you have purchased this book and read this far, we must conclude that you have an interest in the seashore and would like to have it remain in a somewhat natural state. By following a few simple rules you will not only help in this effort, but will also make your visits to the sea more enjoyable and satisfying. The State of California requires that authorization be obtained to remove any animal or plant from our beaches. A California Sport-Fishing License will allow the taking of certain species of game fishes and invertebrates (abalones, clams, crabs, etc.) within the restrictions imposed on sizes, numbers, and seasons. In order to collect **any other organism** you must possess a Scientific Collecting Permit; applications may be obtained from your local Department of Fish and Game Office. To apply for such a permit, you must fully justify your proposed collecting, and subsequently submit a detailed list of all species taken, how many were collected, where they were collected, etc. Since this book may find use in Oregon, the reader should know that the collecting laws in that state are even more restrictive than those in California. It is our hope that most of you will not be collecting things, but rather using this book to study the creatures in the wild; they are much more interesting in their homes than in yours. Unless properly cared for, most specimens quickly die and begin to rot, becoming evil-smelling gelatinous masses to be tossed out with the garbage. Sea water aquaria are extremely difficult to maintain, especially for animals which are accustomed to the chilly waters of the northern California coast. Some collected specimens are not discarded, but may find their way into curio shops and sundry tourist joints. Such lovely items as abalone shells (*Haliotis*) are often adorned with a crucifix which glows in the dark and offers rather questionable lines of communication to the Almighty. Our guess is that He doesn't appreciate those things either.

When you are in the field, remember that the low tides have made the animals and plants especially vulnerable by exposing them to air and to terrestrial intruders. Tread lightly over the rocks and do not leave a trail of squashed beasties in your footsteps. **Always replace overturned stones**! This is the cardinal rule of intertidal conduct. Many organisms can only live either on the top or the bottom of a rock, and leaving their world topsy-turvy means certain death. Large areas have been ruined because this simple practice was ignored.

If you are involved in legal collecting, remove only what is necessary for your work and make full use of your specimens. It is not essential that every child in a class takes home a seastar or a snail. If you are a teacher do not allow (or require as some do) your students to make collections.

It is often tempting for residents of the north coast of California to point to the more populous areas of the state with a "see-what-you've-done" attitude. We have seen it all, and the idea that destruction of the natural environment cannot happen here is misleading and downright dangerous. It can and is occurring even on the most remote of our northern shores. Many local areas have changed over the past ten years as a result of some of the practices mentioned above. Our north coast is still, in many ways, the most beautiful and least disturbed region of the California shoreline. This condition can only be maintained through strict conservation practices, common sense, and self restraint. Otherwise, the marshlands of Humboldt Bay, the pristine beauty of Patrick's Point, and the spectacular expanse of Clam Beach will become things of the past as are so many of the once similar areas to the south.

CHAPTER II

HOW AND WHERE TO LOOK FOR SEASHORE LIFE

Students often tell us that they had visited the shore many times without seeing anything until they had learned how to look with care and patience. Many animals display cryptic coloration, and others are secretive in their habits, especially when the tide is low, so that at first glance the shore may appear as a rather dull array of military greens and browns. One must look closely to see beyond the few large and spectacular animals which dominate the field of vision against this background. (An inexpensive low power hand lens will be very helpful.) Children are often better than adults at finding less obvious creatures because they are built closer to the ground, and their minds are somewhat less cluttered with other matters.

The success of your trip to the shore will depend to a large extent on the state of the tide (see Chapter III). In general, any tide which is predicted at 0.0 ft. or lower will provide excellent exposure of a good variety of creatures. The lower the better, obviously, and tides of -1.5 ft. are not uncommon in the spring and summer months. These seasons also generally provide greater overall abundance of animals and plants than do winter conditions. Tide tables are available at most sporting goods and marine-supply stores.

The types of shoreline environments along this coast can be categorized by the nature of the substrate (rock, sand, mud, etc.) and the degree of exposure to wave action. The details of these and other environmental factors are discussed in Chapter III. Each combination of features produces a distinctive habitat housing its own assemblage of organisms. The techniques necessary to observe the creatures will vary somewhat between these different shore types. The north coast of California is basically a series of rocky headlands separated by sandy beaches and coves of various lengths. This pattern is irregularly interrupted by numerous river mouths and by a few bays, notably Humboldt Bay.

ROCKY SHORES

The variations which exist along rocky coasts are due, in large part, to the action of waves. When approaching such rocky regions it is a good idea to survey the area before actually entering the tidal zone. Incoming waves generally tend to twist and bend to run parallel to the depth contours of the bottom. Thus, when close to shore the wave pattern may change rather drastically. Once your attention is directed to searching among nooks and crannies it is inadvisable to be taken unawares by a breaking wave. Some areas are subjected to full exposure to the surf; others are protected to varying degrees by offshore rocks or the general shape of the coastline. Walk carefully and slowly over the rocks to protect both the sea life and yourself. Many intertidal plants secrete an extremely slippery mucus which helps protect them from drying during low tides. A bad fall can result from moving too quickly or jumping from one rock to another.

Do not limit your search for organisms to any single area of the rocky shore. Start with the lowest exposed areas and cover the entire region up to the highest water marks. Pick carefully through clumps of seaweed and pockets of sand deposited between and under stones. Look under ledges, in crevices, and beneath overhanging rocks. The undersides of rocks will invariably house different organisms from those seen on the upper surfaces. Spend some time sitting quietly beside a tidal pool; your patience will soon be rewarded with a view of an active world ignored by most casual seashore visitors.

SANDY BEACHES

Wave-swept sandy beaches are among the most unstable of all natural environments. The result of this instability (and other factors) is a very low diversity of organisms, but often very high numbers of those types adapted to such conditions. A shovel or other digging tool is generally necessary to uncover most of the sandy-beach animals. Again, you should extend your search from the high-water line down to include the sheet of water left by the receding waves. Dig in the high semi-dry sand and in areas where the surface bears small holes. These openings usually lead to shallow burrows in which some of the sandy-beach animals reside. Other animals may be easily located by turning over piles of seaweed washed up on the beach. Some creatures feed in the wave backwash and may be found by gently sifting through the wet sand just under the water. Certain types of animals and seaweeds normally found offshore may be washed up onto sandy beaches during periods of heavy wave action. A leisurely walk along such a beach following a storm will often yield a variety of organisms not usually seen by the land-bound visitor.

TIDAL FLATS

Humboldt Bay and similar areas provide some of the most exciting intertidal environments along the northern California coast. The absence of continuous wave activity and the gradual slope of the bottom result in the exposure of great areas of sand and mud during periods of low tides. In places, the soft mud can be very difficult to walk through, and the explorer should make sure that his shoes are tightly cinched and should not venture too far alone. More than once an unsuspecting student has required the aid of others with shovels to be extracted from deep mud. The mud and sand flats may appear barren at first glance, and some sort of stout digging tool is a necessary piece of gear for a successful search. Digging by hand may be appropriate in some cases, but beware of broken shells and, unfortunately, glass, imbedded in the sediments. The presence of many tidal-flat animals is betrayed by holes and mounds on the surface (sometimes referred to as the "signatures" of the residents below). Some animals live within a few centimeters of the surface, others may be nearly a meter down in the muck. A bit of practice and accumulation of field notes will soon increase your ability to recognize likely spots in which to dig for particular kinds of organisms. Some animals are associated with the unique plants which live on mud and sand, and others plow along in the surface layer of the sediments. It is wise to search carefully among the delicate algae and especially over the blades of eelgrass (*Zostera*) which forms extensive fields or "beds" on tidal flats. A number of animals are particularly adapted to living on the narrow leaves of eelgrass and exhibit remarkably cryptic coloration patterns.

MISCELLANEOUS AND MAN-MADE HABITATS

Some of the most rewarding places to look for seashore life can be found by searching out some of the unique habitats in Humboldt Bay and other northcoast harbors. Jetties and rock piles partially buried in sand or mud will reveal certain species of animals which are adapted to hard substrates but which cannot tolerate wave action. Floating docks and buoys provide accessible spots to find organisms which normally live below the level of tidal exposure. Wooden pilings, styrofoam floats, and even dockside tires are usually covered with luxurious growths of animals and plants in unique communities not found in natural intertidal habitats. Be certain to request permission to use private areas, and exercise care not to damage piers and other artificial objects. Incidentally, the above applause for man-made structures does not constitute a request on the part of the authors for building more of the same, especially at the expense of natural habitats.

CHAPTER III

III

ECOLOGY OF THE SEASHORE

The term ecology implies a vast and comprehensive overview of life and how its components (communities, species, populations, individuals, etc.) interact with one another and the physical world in which they exist. The breadth of this concept is illustrated by the often-heard remark, "All biological investigations are, in one way or another, ecological inquiries." Obviously, to discuss the ecology of the sea properly would require a text in itself, and indeed, several excellent texts of this nature exist. In this chapter we hope, therefore, to touch only upon the most important and easily observed ecological phenomena of coastlines. Like any human endeavor, science is subject to "fads" and "trends" of various sorts, many of which seem to be cyclical in nature. Somewhat unfortunately, these "fads" of biological interests often tend to reveal themselves in the editorial practices of certain scientific journals, often leading to the exclusion of more "traditional" but nevertheless reliable research reports. In writing this chapter we have kept these popular trends in mind, but we have not given them priority over other aspects of coastal ecology. It is important to remember that while speculation and theory have their place in scientific inquiry, the real foundations of our accumulating body of rational thought and scientific knowledge rest on accurate descriptive and experimental biology, in which these theories and concepts are put to rigorous tests of scrutiny and logic.

This chapter is divided into three sections, each concerned with a different set of factors which influence the lives of seashore organisms. These are physical factors, biological factors, and artificial factors (man-made modifications of the natural environment). This division is, of course, one of convenience, and it must be realized that any or all of the phenomena discussed on the following pages may be operating simultaneously in a given biological system, each influencing the others to varying degrees. Bear in mind, while reading this chapter (and also during your visits to the shore), that animals and plants survive in their habitats through adaptations of three general sorts: anatomical, behavioral, and physiological. Although these adaptations are by no means mutually exclusive, the careful observer will usually notice one or another as apparently dominating the organism's response to any particular environmental situation.

PHYSICAL FACTORS

1. **Substrate Type.** The general biological character of any shoreline is controlled first and foremost by the nature of the substrate. In fact, the three principal habitats of our north coast (indeed, of any temperate coast) may be conveniently classified as rocky shores, sandy beaches, and protected mud and sand areas called tidal flats (figs. 2, 3, 4). The last category includes any enclosed or semi-enclosed coastal region in which wave shock is greatly reduced.

Rocky shores typically harbor the greatest number of kinds (species) of marine organisms. This diversity seems to result from the fact that these rugged coastlines have 1) a large variety of habitats such as crevices, tide pools, drainage channels, and other places where different kinds of animals and plants can live (ecologists refer to the presence of these irregularities in the substrate as **substrate heterogeneity**), and 2) a direct access to nutrient-rich offshore waters. This constant renewal of nutrients plays a dramatic role in the overall biological strength of a locality (as measured by the rate of production of new animal and plant material . . . **productivity**). Both the numbers of kinds and the number of individuals are increased under such conditions. This situation is further enhanced by localized conditions where nearshore **upwellings** bring even greater supplies of nutrients to the surface from deeper water. (Upwellings are risings of deeper, subsurface waters, caused by winds and currents. These

2. Rocky intertidal

3. Wave-swept sandy beach

4. Tidal flats, Humboldt Bay

deep, cold waters are generally rich in nutrients and contain high amounts of dissolved oxygen.) In such coastal regions, which often occur on the western shores of large land masses, particularly around headlands, the localized effects of upwelling may increase productivity several-fold. One of the largest and most well known of such upwelling regions is off the coast of Peru, which (until recent major current changes took place) supported the most productive fisheries in the world.

Most of the rocky-shore organisms live on the surface of the substrate and are referred to as **epibenthic** (**epifauna** for reference to animals only). One can delineate any number of habitats within a rocky shore system: for example boulder habitats, cobble habitats, rock/mud habitats, tidepools, etc. In addition, substrate heterogeneity is further enhanced when the rocks themselves are more porous. The hardness of the rock formations making up the beach and intertidal region influences this substrate porosity or texture (referred to as **rugosity**) and hence the types and numbers of organisms that can inhabit any particular shore. In general, the harder the rock, the less rugosity it has. Beaches composed of hard smooth rocks, like basalt and diabase, are often found to have fewer kinds of animals (i.e. a lower **species richness**) than shore lines composed of softer sandstone, limestone, or granite. Rocks of low rugosity simply do not offer as many cracks, holes, and depressions which animals can use as homesites. In addition, the surfaces of smooth dark rocks will reach a considerably higher temperature during daytime tidal exposures than will the surfaces of lighter-colored, more rugose rocks. These variables will clearly influence the types of animals and plants that can exist on these substrates. A short time observing these phenomena in the field will quickly reveal to the burgeoning naturalist just how complex these fundamental concepts may be.

Still another aspect of rocky shores, and any other kind of natural shoreline for that matter, is that of **substrate stability**. Shores composed of very large boulders are physically more stable than shores composed of smaller rocks, loose stones, or cobble. One can see a direct correlation on these shores between substrate stability and species richness. Loose cobble areas, such as Agate Beach near Patrick's Point, are rearranged during each period of heavy wave action, and very few species of animals are capable of existing on such a shore; the homes and bodies of most would be tossed about and quickly ground between the stones on such an unstable beach. These cobble and shingle beaches are nearly always derived from areas which are subjected to high amounts of erosion and subsequently reworked by wave action.

The shores with the largest numbers of species are usually ones with a variety of substrates (high heterogeneity), with porous rocks (high rugosity), and with large erosion-resistant rocks (high stability). A mixture of large and small boulders, semi-cemented together by both sand and mud, approximates such an environment.

Trinidad, Luffenholtz, and Endert's Beach Cove, among others, offer these conditions.

In contrast to the rocky shores, sandy beaches typically harbor the fewest kinds of animals and plants of any coastal habitat (except perhaps a small-cobble or shingle beach). Two important factors apparently contribute to this low species richness: (1) both substrate heterogeneity and stability are very low on sandy beaches, and (2) coarse-grained sandy beaches tend to drain better and faster than rocky shores, fine sands, or tidal flats, and a great portion of the beach literally "dries up" during each low-tide period. The only macroscopic animals present are those that are tough enough and/or agile enough to withstand the scouring effects of the constantly moving sand and migrate to maintain a position where moisture is nearly always present. Four classic examples from our coast are: the beach hoppers (amphipods, fig. 53); the mole crabs (*Emerita*, fig. 65); the razor clams (*Siliqua*, fig. 117); and a cirolanid isopod (*Excirolana*). Each of these animals is capable of maintaining a more or less constant position relative to the tide level. By migrating up and down the beach slope, or moving vertically through the sand, these creatures stay in areas of adequate moisture. Exactly how such complex and consistent behaviors are controlled is not fully understood.

One advantage of occupying such a rigorous environment as a sandy beach is that living space and food resources are shared with very few competitors. As a result, those animals which can survive on such a beach usually occur in very high numbers and dominate large areas of the beach.

There is another world of animals and plants inhabiting sandy beaches that can only be briefly mentioned here, the minute creatures which live amongst the grains of sand themselves. These **interstitial** organisms are usually less than a millimeter or so in length and include creatures of various sorts whose lives are spent crawling about and between sand grains. This "world in miniature" has only recently been investigated by biologists, and no doubt only a small fraction of the species that exist there have so far been described and named. Most of this interstitial fauna is composed of various worms (nematodes, annelids, etc.), minuscule crustaceans (mystacocarids and copepods), protozoans, and certain odd, very tiny sea anemones.

Tidal flats are typically formed in soft-bottomed, shallow, and gently sloping coastal regions, such as bays and sloughs. Because of the gentle slope of these shores, the area exposed by the receding tides (i.e. the breadth of the intertidal zone) is usually much greater than that on steeper open rocky coasts and sandy beaches. In addition, the protected nature of a bay produces significantly less wave shock and surge. The combination of these two factors, plus the fact that these regions often have sediment-laden rivers emptying into them, generate a muddy or sandy-mud bottom, which harbors entirely different kinds of animals and plants from those seen in any other coastal habitat. Whereas the hallmark of success on a rocky shore is the ability to avoid being washed away or crushed by the constant pounding of the waves, most of the animals inhabiting a tidal flat have conquered this habitat by using highly efficient burrowing techniques. It is for this reason that a tidal flat, when first viewed at a low tide, usually appears as bleak and barren as a sandy beach. In fact, it lies somewhere between the sandy beach and the rocky shore in terms of species richness, productivity, and overall **biomass** (i.e. the amount of living material in the area). It is simply that the vast majority of the inhabitants of a tidal flat are members of the **infauna**, that is, they live beneath the surface of the substrate and therefore are hidden from view.

Much of our basic knowledge concerning life in these soft-bottom habitats, both between the tides and below the lowest tide mark, stems from the pioneering work of the Danish naturalists who found the fjords of Denmark ideal natural laboratories for the study of such environments. Through the patient efforts of the Danes, and a small number of fine biologists that have followed them, we now know that benthic animals living in soft bottoms are not at all randomly distributed, but tend to be found in distinct associations. The composition of these associations, or **communities**, appears to be strongly correlated with: 1) grain size of the sediments, 2) amount of organic matter in the substrates, and 3) tide level.

The inhabitants of a tidal flat feed primarily on particulate organic matter (**detritus**) that they either filter from the water overhead or extract from the sediment itself. These animals include the clams, many snails, certain burrowing sea anemones, the ghost shrimps, and a great variety of worms (polychaetes, sipunculans, echiurans, etc.). There are, of course, some epibenthic predators that crawl about the mud surface in search of prey. The most obvious, on our north coast, is the voracious moon snail (*Polinices*, fig. 98), which feeds primarily upon other snails and clams (discussed later).

If one were to sift some sediment taken from a tidal flat through a set of sieves, each of a different mesh size, one would see that a number of different grain sizes make up the sediment. These generally range from large shell fragments, through large and small grained sands, to fine silts and clays. Ecologists are quite concerned with the composition of sediments in such regions, because the percentage of each grain size in a particular place will greatly influence the physical nature of the habitat and hence the kinds of animals that can live in it. Very sandy soils, for example, drain rapidly but are generally well oxygenated, while very silty soils (fine muds) usually retain moisture throughout a low tide exposure but often are low in available oxygen because of slow rates of internal circulation and other factors. One does not find a uniform composition of sediments throughout a tidal flat. The reason for this diversity is that currents (and to a lesser extent the digging activities of various animals) tend to distribute the sediments, so that patches of very sandy substrate are found in one region of the bay, while in another area the sediments might be very clay-like or silty. The kinds of animals and plants associated with these different patches of substrates vary from one to the next. All other factors being equal, a tidal flat composed of many types of sediments (i.e. having a high substrate heterogeneity) will harbor more species of animals than one with only a few different sediment types. Further, should such an area have rocky outcroppings in addition to the usual soft substrates, species richness is again significantly increased.

The foundation of any food chain consists of the photosynthesizing plants that convert water, carbon dioxide, and the sun's energy into carbohydrates. The basis of the food chains in an open-coast rocky shore, or in the open sea, is largely the microscopic and macroscopic plants that live there (primarily algae). On a tidal flat, however, plant production is often very low, and the basis of the food chain is shifted from living material largely to detritus. The small particles of organic material making up the detritus are the residue from dead plants and animals. The detritus comes from the open coast, river and stream drainage, adjacent salt marshes, and to a lesser extent from the tidal flat itself. The point is, few tidal flats or bays are self-sustaining, and most must rely on outside detrital sources to provide the energy input that supports their animal populations. There are exceptions, of course. Some tidal flats harbor luxuriant stands of eelgrass (*Zostera marina*), which can provide an enormous energy input, both directly to the grazing herbivores, and indirectly to the **detritivores** (those animals which feed on detritus). In addition, some bays, such as Humboldt Bay, and estuaries support extensive growths of certain semi-marine plants, such as glasswort (*Salicornia*), cordgrass (*Spartina*), and others. Although very few marine animals feed to any significant extent on these shoreline plants when they are living, the plants do contribute considerable vegetable detritus to the food chain of the bay. In these cases, the tidal flat may become one of exceptionally high productivity, largely because of the contributions of one (or only a few) species of nearby plants. In such a system ecologists occasionally refer to the primary producing level (the plants) of the food chain as being dominated by a **monoculture**, that is, one species. Although these tidal-flat systems never approach the level of species richness seen on a stable rocky coast, a high-energy (highly productive) tidal flat will usually have a considerably greater number of individual animals and a greater number of species inhabiting it than will such an environment in which healthy growths of macroscopic plants are lacking.

In cases where there is simply too much detritus coming into the system for the animals to consume, bacteria take over. These little creatures multiply to vast numbers and can quickly deplete the sediment of its available oxygen. When this occurs other types of bacteria become abundant, ones that utilize the element sulfur instead of oxygen for their normal metabolic processes. The presence of such sulfur bacteria is easily detected because they typically release hydrogen sulfide (H_2S) as a metabolic byproduct (instead of water,

H_2O, like regular bacteria, plants, and animals). This hydrogen sulfide combines with iron in the sediment to produce a dark black layer in the mud (the black color being iron sulfide). This black mud is usually located a few centimeters beneath the surface, particularly in fine sediments, at the depth to which oxygen does not readily circulate. Even the most casual observer will notice the potent aroma of hydrogen sulfide (also known as "rotten egg gas") that accompanies these anoxic (without oxygen) conditions. Organisms which live buried in such sediments must either be able to tolerate very low oxygen levels or have some mechanism of obtaining oxygen from the water above the substrate. Many clams, for example, reach above the sediments by means of fleshy tubes (called siphons, fig. 115) through which they draw in and expell water. Do not let the smells or soft sediments of the tidal flats discourage you, however, for in these unique habitats you will find magnificent and unusual animals that occur nowhere else on our coast.

2. **Wave Shock and Currents.** The most obvious features of the crashing waves of our Pacific shores are the tireless, unrelenting consistency with which they pound against the beach substrates, and the enormous amounts of energy released as they do so. Obviously, one of the first and foremost prerequisites for life on an exposed outer coast is the ability either to "hang on" in the face of these forces, or avoid the situation entirely by hiding beneath some suitable portion of the habitat (i.e. under a large rock, deep within a mass of seaweed, etc.). We offer below just a few examples of how animals have evolved to facilitate these two strategies of dealing with the surf. More information on such adaptations is included in the discussion on the particular animals, in later chapters.

Perhaps the undisputed champions of the "hangers-on" are the limpets. These tenacious little snails have evolved simple uncoiled shells that resemble flattened dunce caps (figs. 80-89). Their large muscular foot is capable of clamping down on a rock with a great deal of force, while their streamlined shell serves to deflect the forces of a crashing wave down its smooth sides to the rock surface below. Many animals of the exposed rocky shore have similar structures and/or clinging abilities (i.e. barnacles, abalones, mussels, etc.).

Some of the best examples of the "hiders" are the various worms and worm-like animals, which are often soft-bodied and lack a protective covering or a means of firm attachment. Peanut worms (sipunculans, fig. 43) that inhabit such regions are invariably found hidden away beneath large rocks or within the protection of a mussel bed. There are even a few species of tropical sipunculans that are actually capable of burrowing directly into soft rock, thus avoiding the detrimental forces of the pounding surf while at the same time enjoying the benefits of the circulating sea water. Most polychaetes (segmented worms, figs. 35-42) are found in similar abodes, the only ones capable of living exposed to waves being those that produce hard protective tubes which are firmly cemented to the rock surface. A few minutes spent observing the animals of these surf-exposed habitats will quickly reveal their myriad range of adaptations to living in such an environment. Nevertheless, the pounding surf does take its toll. The aged, injured, or infirm which cannot cling tightly enough eventually become dislodged or abraded from the rocks to become food for the scavengers and detritivores. On our north coast, where driftwood is common, a crashing wave often carries with it a natural battering ram (logs lost from lumber rafts are especially potent in this regard). Such heavy pieces of floating debris are capable of not only smashing or maiming the toughest intertidal organism, but they often tear great portions of mussel beds off of rocks. This latter form of damage is especially serious in older mussel beds, in which the uppermost members are living upon a layer of dead shells and thus attached relatively weakly to the substrate.

Waves are not all bad, of course, for they help to keep the water well aerated, as well as keeping minute particles of food (detritus, etc.) in suspension. In addition, the particular patterns of water flow and surge are utilized by many animals to assist them directly in their feeding activities. Barnacles and mole crabs, for instance, can both be seen to "face" into the predominating water flow (usually a backwash), whereupon they can simply open their filtering devices and allow the energy of the moving water to do the work of filling them with food.

Waves can also play a vital role in restructuring our coastlines, as they constantly erode the shore away and thereby open up new habitats which animals and plants can then exploit. Storms typically erode sandy beaches each winter, creating steep shore profiles as they carry tons of sand offshore into deeper, calmer waters. Summer surf, being of a gentler nature, then moves this sand back on shore to create gently sloping beaches. This ability of moving water to transport sand, stones, boulders, animals, etc. is primarily a factor of its velocity at any moment in time, and in fact, varies as the sixth power of the velocity. This means that if a rock one cubic inch in size is carried about by a wave of some velocity, a stone of 64 cubic inches can be carried by a wave twice that velocity. Once one understands this mathematical relationship it is not difficult to understand the ability of storm waves to damage breakwaters. For example, a breaking wave falling through a distance of about four feet (not a particularly large wave for our coast) strikes with a velocity of about 16 feet per second and can move a 64-cubic-inch rock; while a wave falling 16 feet can move a boulder of nearly 40 cubic feet! It is no wonder that so many plants and animals are torn off of rocks during winter storms.

Nearshore currents are vitally important to coastal animals and plants. They are clearly responsible for distributing the nutrients in the sea from one place to another. This distributing nature of currents is vital to coastal life in another way. The majority of the animals and plants of our coast release their eggs or larval stages freely into the water, relying upon the currents to distribute these products of their reproductive efforts up and down the coast. It is by this means that the ranges of these organisms are maintained or expanded.

The southward-flowing waters of our coast form the well-known **California Current** and are fairly cold, partly because of the large amount of upwelling which occurs in this region, particularly in the summer months, and partly because of their origin in high latitudes. In winter and early spring, however, a deep-water current flowing northward usually rises to the surface near shore. This relatively warm current from the south is called the **Davidson Current.** Its effects are to push the cool waters of the California Current offshore somewhat, and keep our coastal winters relatively mild. In addition, in a year during which the Davidson Current is well developed, many animals with a more southern distribution will have their larvae carried northward where they may settle on our northern shores.

3. **Temperature and Oxygen.** Water temperature is probably the most important single factor influencing animal and plant distributions in the sea. Every species has a temperature range in which it can survive and maintain healthy, reproducing populations (this range is often referred to as the **Zone of Activity**). Above and below this range an animal or plant is subjected to stress and, although an individual may survive outside its optimal thermal limits, it cannot compete adequately to obtain the food and shelter it needs, nor can it maintain successful reproductive activities.

As one moves south along the coast of western North America, from the Arctic regions of Alaska towards the equator, one encounters a gradual trend of increasing temperature regimes. North of the Bering Sea, where sea ice is encountered throughout the year, is the **Frigid Region** (water temperatures near or below 0°C year around); most of the Pacific northwest coast (including California as far south as Point Conception near Santa Barbara) belongs to a fairly-cold-water regime called the **Cold Temperate Region** (water temperatures on open coasts typically fall below 10°C in the winter, while summer means [averages] may approach 20°C); south of Point Conception, to the west coast of Baja California, is the **Warm Temperate Region** (water temperatures average between 10°C and 20°C in the winter, with summer maxima around 25°C); and south of this lie the subtropics and tropics. The mean summer water temperatures along our own portion of the northern California coastline are in the range of 11-14°C, while winter temperatures average around 8-10°C.

As you have no doubt guessed by now, each of these regions contains characteristic assemblages of marine plant and animal species. Although a small degree of overlap certainly occurs between adjacent areas, most animal species have their ranges confined to only one (or two) of these particular regions. When these regions are defined on the basis of their plant and animal communities (instead of temperature) they are referred to as **biogeographic provinces.** Although the exact boundaries of such biogeographic provinces are dependent upon which groups of animals or plants one analyzes, they generally conform to the temperature regions described above. Biologists often recognize various numbers of subgroups within each of these categories. Our north-coast seashore creatures are inhabitants of the so-called Cold Temperate Region or **Oregonian Province.**

One interesting exception to this concept of restricted biogeographic distribution is the phenomenon of **submergence.** Just as water temperatures drop markedly as one moves northward along our coast, they also drop markedly as one moves offshore into greater depths. Because of this, a number of north coast animals have been able to extend their ranges southward by moving into deeper water as they do so, thus maintaining themselves in temperature conditions similar to those which the more shorebound northern populations experience. It is for this reason that bottom trawlers or fishermen will occasionally capture a typically cold-water animal far south of its "normal" distribution or range. By the same token, many of the southern Channel Islands, around which upwelling frequently occurs, will usually harbor a complex mixture of seashore life related not only to the local nearshore fauna, but also to the fauna of more northern coasts.

The effects of water temperature relate not only directly to the animal's ability to function in a normal manner, but also indirectly by influencing other environmental features. For example, the amount of dissolved oxygen that water can hold is a function of its temperature; the colder the water the more oxygen it can retain. Thus we find that the cold waters of our upwelling regions provide coastal sea life not only with an increased nutrient availability, but also with an increased oxygen concentration. Similarly, a high intertidal pool, left stranded in the summer sun, may occasionally become so warm that the oxygen level in the water drops to near zero.

The organisms with which we are primarily concerned in this book may be found within the intertidal zone, on that portion of the shore that is alternately covered and exposed by the ebb and flow of the tides. Clearly, these organisms are confronted with more than just the relatively stable temperature conditions of the sea itself. Indeed, most are exposed daily to a variety of temperature conditions, which include not only aerial temperatures, but also the greatly depressed temperatures which are produced by winds and evaporative cooling. Alternately, animals stranded in a small tidepool on a warm summer afternoon may experience an enormous temperature increase as the sun heats the water in the pool. Needless to say, those members of the sessile or slow-moving epifauna, and attached algae, that are exposed to these fluctuating conditions have developed a great deal of resistance to such short-term temperature and oxygen extremes. They have also evolved a number of simple but effective mechanisms for ameliorating the potentially lethal effects of these conditions. Many snails and tube-dwelling worms, for example, close the opening of their shell or tube during the low tide hours, sealing it shut with a mucus-like "glue" and a horny door (called an **operculum**). Other animals trap small amounts of water within their burrow or shell; some, like the limpets and abalones, may clamp down against the substrate so tightly that they form a nearly air-tight seal against desiccation; many algae secrete copious amounts of slime or mucus that serve to keep their bodies coated and moist during periods of exposure to air.

Obviously, those organisms that live highest in the intertidal zone are exposed to non-aquatic conditions the greatest number of hours each month. Some of the high intertidal snails known as periwinkles (*Littorina*, figs. 90-91), and barnacles (fig. 47) are actually so well adapted to these long exposures that they have been kept for many weeks completely out of contact with water (on a biologist's desk-top or other such unlikely place), to revive completely, in perfect health, upon their return to the sea. Some of these creatures, in fact, cannot tolerate prolonged periods of submersion and must spend most of their lives out of water to survive.

4. **Salinity**. Salinity is defined as the concentration of salt(s) in water. The salinity of the open sea is very stable, varying only a few tenths of a percent, from about 3.3 to 3.7 percent in the world's oceans (this salt concentration is usually expressed in parts per thousand; 33 to 37 ppt.). This small degree of variation probably presents no particular problems to any marine organism. There are, however, three principal situations in which variations in salinity do play a significant role in the life of a marine animal or plant. One occurs when a high tidepool is left stranded for many hours exposed to the warm summer sun. This tidepool may experience not only a significant rise in temperature, but enough evaporation to raise the salinity considerably. In the subtropics and tropics salinity in such a tidepool may double during a five- or six-hour period of exposure (to 7 or 8 percent), a change which is definitely stressful for creatures not well adapted to such extremes. On our cool, north coast, however, such drastic increases in tidepool salinities are quite rare, although salinities of 4 to 4.5 percent are not uncommon in high isolated pools on sunny days. Most of the animals that inhabit these high intertidal pools are well adapted to handle these changes, just as they can handle the increase in water temperature. An animal taken from the lower tide levels, on the other hand, may not be able to tolerate such rapid shifts in temperature and salinity and could die if moved to this different habitat.

A second place in which salinity is of concern to marine life is in or near the mouth of a river. A large river will push a tongue of fresh water a considerable distance into the sea. One of the classic examples often cited of this phenomenon is that of the Amazon River, on the coast of Brazil. It has been said that sailors of old were able to obtain their first taste of pure fresh water after a long journey across the Atlantic when they were still a mile off of the mouth of the Amazon. Although none of the rivers on our north coast approaches this magnitude, the effect of their freshwater runoff is certainly significant, if localized. Fresh water tends to float for a considerable length of time on salt water, because it is lighter (containing far less salt than sea water). Thus, in an estuary one will often find a **salt-water wedge** lying on the bottom with the thin end of the wedge pointing upstream. One may find, in fact, fresh-water organisms living in the surface waters of an estuary, while a distinctly marine fauna flourishes in deeper water and on the bottom. The degree to which the integrity of this fresh-water/salt-water layering is maintained depends upon the amount of water turbulence in the estuary. Naturally, if the region is fairly exposed to waves, or subjected to large tidal fluctuations or strong currents, a greater amount of mixing will take place and the stratification will break down. In these situations, a well-developed **salinity gradient** may form, instead of a salt-water wedge with a distinct boundary. That is, the salinity will be seen to increase rather evenly and gradually towards the mouth, from fresh, through brackish, to sea water.

A third situation in which salinity variations occur is, of course, during periods of heavy rainfall. Isolated tidal pools may display significant salinity reductions when the sea water is diluted by rain. Unless this condition is compounded by heavy local runoff, however, tidepool organisms are usually only exposed to these lowered salinities for a short period of time and upon the rising of the tide are bathed in water of a more appropriate salinity. The combined effects of heavy rain and increased river runoff may present serious problems to intertidal life. Most marine animals cannot live if the salinity drops much below 2.0 percent (20 ppt.), especially if such conditions persist for more than a very few hours. It is for this reason that in the rainy season a certain amount of localized die-off of plants and animals may occur around the head of an estuary. Populations thus affected, however, usually recover quickly once the river flow has returned to normal.

There are very few animals or plants that are capable of surviving in both fresh and salt water (the various species of salmon are classic examples of these exceptional animals, and even these require periods of adjustment when migrating from one condition to the other). This inability of most animals to survive in both fresh and salt waters has one distinct advantage for man. With only a few recently discovered exceptions, there are practically no freshwater pathogenic (disease-causing) organisms that can thrive in the sea. Considering the enormous amounts of partially treated (and untreated) sewage that we dump onto our coastline every day, this natural sterilizing effect of the oceans is truly a blessing. Residents of the Great Lakes region are, unfortunately, not so lucky, and instead rely on the outdated and dangerous philosophy that "dilution is the solution to pollution," to handle their lake-bound sewage.

5. **The Tides and Intertidal Zonation:** No attempts will be made here to explain the details of the physics and astrophysics of tides. The dynamics of these phenomena have been amply explained in a great number of texts, and the determined reader is referred to the references at the end of this book for information beyond what is given in the following passages. Perhaps the most lucid lay account on this subject is that found in Bascom's excellent little book, **Waves and Beaches** (1964).

The tidal flow of our north coast is generally described as **semidiurnal** (or **semidaily**). These terms refer to a tidal pattern of two high tides and two low tides each day. (Because of the strong influence of the moon on the tides, the observable tidal cycles are based upon a lunar rather than a solar day. Thus a particular point on a tidal cycle will occur about 50 minutes later on each successive day by our clocks.) In certain areas of the world only one high and one low tide occur each day. These patterns of highs and lows are associated with, and largely dependent upon, the relative positions of the moon and sun with respect to the Earth (the phases of the moon), and the height of the moon in relation to the equator. The tides are also influenced to a lesser degree by the centrifugal force of the Earth-moon system around the sun, topography of the ocean basins, and that phenomenon known as **seiche**, or the natural period of oscillation of large bodies of water. During the new and full moons the Earth, moon, and sun are in an approximately straight line (they are said to be "in phase"), and the tidal range is at its greatest because of the multiplying effect of the gravitational pull of these two celestial bodies on the Earth's seas. These extreme high tides and extreme low tides are referred to as **spring tides** (not because they occur in the spring, but because the coastal waters seem to "spring" back or well up on the beach after the low tide). During the quarters of the moon, when the sun, Earth, and moon form a right angle, the tidal range is least, falling 10 to 30 percent below the average range; these least extensive tides are known as **neap tides.**

Unlike the northeastern coast of North America, whose two daily highs and two daily lows are approximately equal, most of the California coastline experiences **irregular** or **mixed** semidiurnal tides, in which one daily high is much higher than the other high, and one daily low is much lower than the other low. The degree of this irregularity varies at different coastal sites, largely because of differences in local topography.

So, we have two low tides and two high tides occurring each day, which twice a month (during the full and new moons) develop into very high highs and very low lows. The breadth of the beach that extends from the highest point the high tides cover, to the lowest point the low tides expose, is known as the **littoral** or **intertidal zone.** It is with this area that this book is primarily concerned. Below the low-tide line lies the **subtidal zone,** accessible only by wading, diving, or trawling. Above the high-tide line, in the narrow band where the winds and waves toss their salty spray, is the **splash** or **spray zone.**

Above the spray zone is the terrestrial environment, a land dominated (in terms of numbers) by the insects and grasses.

For the sake of convenience, the intertidal zone, or more correctly the true tidal range, is divided quantitatively into a number of arbitrary divisions. In North America the unit of measurement used to divide the intertidal zone is the foot (although with this country's conversion to the metric system, this unit will soon become the meter; Appendix 1 lists commonly-used conversions between the two systems). For example, in Trinidad Bay, California, the lowest low tide of the year is separated from the highest tide of the year by about 10 feet of vertical displacement. In other words, if a pole were fixed in the water and observed daily, it would be found that the water level on the pole moves up and down about 10 feet throughout the year, between its highest and lowest points. In order to allow for long-term fluctuations in the level of the sea, we do not start measuring with 0 at the lowest point on the pole, or the lowest low tide, but rather compute a 0 point at some defined level between the extremes and call everything above this level a plus tide (e.g., +2 feet) and everything below it a minus tide (e.g., −2 feet). The actual physical point of the 0 tide level may fluctuate from decade to decade as the level of the seas rises or falls and new calculations are made. The 0 tide level or "0 datum" on U.S. tide charts is the mean (average) of the lower of the 2 low tides each day of the month (defined as **mean lower low water**), thus incorporating neap-tide values as well as springs. We might mention that this system differs from that of Latin America, where the 0-datum on tidal charts is usually set at the mean height of the lowest water of the spring tides only (i.e., mean lower low water during the spring tides throughout the year). As a result of this difference in 0-datum, the 0 tidal level in California is roughly equivalent to the +1-foot tidal level on Mexican shores and those of Central and South America (as based on those countries' tide charts). Mean sea level on our coast is at approximately the +3-foot tidal level, and at about the +4-foot level in Latin America.

The vertical displacement of water along the coast of California varies from about 6 to about 12 feet. For the sake of comparison, tidal fluctuations on most of the eastern seaboard are around 2 to 3 feet; in the northern Gulf of California around 20 to 30 feet; and, in the Bay of Fundy (having one of the most extreme tidal ranges in the world) they occasionally exceed 50 feet. It must be remembered that these figures are for vertical displacement, up and down the pole we fixed in the sea bottom. When these vertical displacements are translated to horizontal displacements the figures vary with the slope of the beach. On a northern California rocky shore with a rather steep slope the tide may "go out" 50 to 100 feet, while in a gently sloping bay or on a tidal flat the water may recede several hundred feet.

Probably since the first naturalist took the time to observe the seashore carefully, it has been apparent that the intertidal zone could be divided not only quantitatively by increments on a measuring stick, but qualitatively, by observing the organisms living in this area. Even a casual observer strolling down an exposed stretch of rocky intertidal beach will see these bands of organisms (called **zones**) occurring one below the other. Ed Ricketts in his original studies of intertidal zonation on the Pacific Coast of California recognized four zones or regions in the intertidal. This original scheme, and numerous others, have been discussed and compared by Hedgpeth in his latest revision of Ricketts' book **Between Pacific Tides** (see references).

For convenience we will consider the shore to be divisible into the following biologically recognizable zones, modified somewhat from the schemes presented in **Between Pacific Tides.** The **spray zone** includes that area which is virtually never covered by high tides, but which receives spray; obviously the width of this zone will vary greatly depending upon the amount of local wave activity and the time of the year. The **high-tide zone**, along most of our north coast, may be considered as extending from the level of the highest tides down to approximately +5 feet on the tidal scale; this zone is uncovered at nearly every low tide. A broad **mid-tide zone** ranges from about +5 feet to +1 foot and may be subdivided into upper-mid, mid-, and lower-mid- regions for more specific designation of levels. This mid-tide zone is completely uncovered during the spring low tides, and at least partially uncovered by most other lows. Finally we may think of that area from approximately +1 foot down to the level of the very lowest tides as the **low-tide zone.** While these numerical definitions are, in a sense, arbitrary, they offer a set of reference points which can be used in describing where various organisms live on the shore. The biological zones as observed in the field are real, even though they may not always fit precisely the increments of our above definitions, especially in the lower zones. For example, most of the animals and plants living in the low-tide zone can also be found living below the intertidal to depths of 20 meters or so. Also, variations in local conditions, such as wave action, drainage rates during ebb tides, etc., may change the vertical ranges of organisms.

At this point it is worthwhile to list a few of the very common animals and plants characteristic of the above-described zones on rocky shores; all of these species are discussed later in this book.

Spray Zone:	*Ligia* (rock louse, figs. 54-55)
	Littorina (Periwinkles, figs. 90-91)
High-Tide Zone:	*Endocladia muricata* (tufted red alga)
	Chthamalus dalli (acorn barnacle)
	Balanus glandula (acorn barnacle, fig. 47)
	Collisella digitalis (High-tide-zone limpet, fig. 81)
Mid-Tide Zone:	*Pelvetiopsis limitata* (rockweed, fig. 160)

Fucus distichus (broad-bladed rockweed, fig. 159)
Nucella (rock snails, figs. 93-94)
Tegula funebralis (black turban snail, fig. 92)
Pachygrapsus crassipes (lined shore crab, fig. 70)
Mytilus californianus (California mussel, fig. 110)
Pollicipes polymerus (goose barnacle, fig. 49)
Pisaster ochraceus (ochre star, fig. 128)
Anthopleura elegantissima (aggregating anemone, fig. 26)

Low-Tide Zone:
Egregia menziesii (feather-boa kelp, fig. 163)
Phyllospadix (surf grass, a good indicator of the 0 tide level)
Hemigrapsus nudus (purple shore crab, fig. 71)
Anthopleura xanthogrammica (giant green anemone, fig. 27)
Pycnopodia helianthoides (many-rayed seastar, fig. 126)
Strongylocentrotus purpuratus (purple sea urchin, fig. 134)

The vertical distribution of all littoral animals and plants is primarily limited by the availability of their preferred substrates. Within their basic substrate requirements one often finds that the upper limit of their distribution is set by abiotic factors, such as desiccation and extreme temperatures associated with exposure times. We have already seen that certain organisms have mechanisms for withstanding longer periods of exposure to air than others. The lower limit of many intertidal animals, on the other hand, is often established by the biotic factors of competition (for food or space) and predation. The sharpest zonation patterns are usually produced by these biological interactions between species, some examples of which are included in the discussion below.

BIOLOGICAL FACTORS

1. **The Concept of Species Diversity.** It is obvious that different organisms which live in the intertidal zone interact to various degrees and in a variety of ways. It is important to be able to evaluate the effects of these interactions on individual species, and on the community as a whole. The nature and extent of the effects are related to the number of different species and the numbers of each species inhabiting an area, and some method of analyzing this diversity is critical to an understanding of the ecology of any environment. If one were to remove every single animal from a specific habitat, for instance a tidepool, and examine every one which had dwelled therein, a number of significant facts would be evident. For one thing, a certain number of species would have been inhabiting the pool. Let us say, for the sake of example, that we find 40 species of animals, a fairly realistic count for a large tidal pool on our coast. This number is referred to by ecologists as the **species abundance** or **species richness** of that particular habitat. Now, if we were to count how many individuals of each species were present in our sample, we would find that some were much more numerous than others. Just how evenly the numbers of individuals of each species are distributed through our sample is referred to as the **species equitability** or simply the **evenness** of the sample. Those species which are found to be exceedingly common, dominating the sample, are usually said to be the **dominant species.** The concept of dominance is in itself a rather complex notion, and we will mention here only that one can measure dominance in a variety of ways, for example: using numbers of individuals (as above), using overall biomass of each species (the combined weight of the individuals of each species), using the amount of food each species presumably consumes, etc.

Many ecologists are very concerned with the measurement of these two factors (how many species are in a particular habitat, and how evenly they are represented there). It would appear that the relative values of these measurements, when compared to such measurements in other habitats and in other geographic localities, can shed light on some very important and fundamental ecological concepts. These measurements are so important, in fact, that a great variety of mathematical equations has been devised to express the relationship between species richness and species evenness. Some of these equations tend to stress the importance of richness, while others favor evenness. Which equation is used to describe a particular habitat is a matter of continuing debate among ecologists. Collectively, these mathematical equations are said to be measuring, or estimating, the **species diversity** of the area which is sampled. One may calculate species diversity of shrimp in tidepools at various localities, or of crustaceans in different tidepools, or of all animals combined in different tidepools. One can even measure the species richness and species diversity of producers (plants), herbivores, and carnivores separately in different habitats.

The flourish of interest in measuring species diversity during the past two decades has enabled us to quantify some remarkable but consistent trends in animal and plant ecology and biogeography. For example, one of the most striking of these biogeographic patterns is a strong trend towards increasing species richness (and generally diversity) as one moves from the higher latitudes towards the equator. This means that, in comparable habitats, one would generally find more species present the further south one moved (in the Northern Hemisphere). Thus, a low intertidal pool on a rocky shore in western Mexico would probably contain a higher species richness and diversity than a comparable pool in Humboldt County. The tropics are, generally speaking, the most diverse of all regions, and a rocky tidepool in Ecuador (on the equator) would probably have an even higher diversity than the Mexican habitat.

Observations comparing similar habitats in different areas have also indicated that diversity is higher in relatively stable or predictable habitats than in unstable or unpredictable environments. Hence, one would find more animals and plants in a tidepool at the 0 tide level than in one at the +4-foot tide level (where exposure time to aerial conditions is greater). We saw another example of this general phenomenon when we compared unstable sandy beaches to stable rocky shores, and cobble beaches to large rocky outcroppings. To complicate this idea that stability enhances high diversity, it should be mentioned that some recent thoughts on the subject suggest that, within a reasonably stable environment, moderate amounts of small-scale disturbance actually help to maintain a high diversity by preventing overall dominance by one or just a few species. Things are never as simple as they seem.

Still another cause postulated to account for increased species diversity in certain habitats is the presence of generalized predators, one that feeds more or less indiscriminately on whatever it encounters. According to this hypothesis, if our tidepool had no generalized predators, eventually a few species would tend to dominate the habitat. The reasoning here is that certain animals in that tidepool simply grow faster and reproduce more quickly, and hence tend to "outgrow" or "out-compete" the others in one way or another. However, should a generalized predator be introduced into the pool, simple probability says that such an animal will be most frequently eating whatever is most common. Hence, these predators serve to keep any one particular species from becoming "too" abundant — the result is more room for everybody and an increase in species diversity. This phenomenon may be seen in action if one observes a large mussel bed on a rocky shore. *Mytilus californianus* (fig. 110) is an animal that grows rapidly and can quickly "take over" the substrate, to the exclusion of many other animals and plants. The ochre seastar *(Pisaster ochraceus,* fig. 128) is a predator which feeds on a large number of different invertebrates including *Mytilus.* If it were not for *Pisaster,* local mussel beds would be much larger than they are and would dominate a much greater portion of the rocky-shore substrate (to the exclusion of other species of animals and plants).

Another example which illustrates the effect of non-specific predators is the sea otter *(Enhydra lutris)*. This animal feeds on a variety of invertebrates including crabs, abalones, and sea urchins. Where sea otters are present, the populations of the prey species are all kept more or less in a balance and no one species dominates the environment. The otters feed most frequently on sea urchins, since they are the most numerous prey, which results in more abundant and dense growths of large algae (kelps which are normally eaten by the sea urchins). The increase in the amount of algae supports more species and more individuals of various invertebrates, enhances the available shelter and food for many fishes, and these support certain birds, seals, sea lions, etc. By the same token, the great reduction in the number of sea otters (see Chapter VII) has had long-lasting and far-reaching reverse effects. The absence of this predator from many regions has allowed the urchin populations to increase tremendously. The decrease in the amount of large algae resulting from overgrazing by these urchins has caused related decreases in species richness among invertebrates, fishes, and so on into higher levels of the food chain.

As one becomes increasingly familiar with the seashore life of our north coast, these ecological trends will become more obvious, and the committed seashore buff may wish to explore an even greater variety of habitats, comparing one with another in terms of species abundance, evenness, dominance, stability, and the presence of generalized predators. All of these things can be observed with little or no disturbance to the habitat in question. Comparisons of the sessile or sedentary fauna living on rock surfaces in different localities, for example, are easily made; but, be sure to take along this handbook and a pencil and pad for recording your field notes.

2. **Competition and Predation.** Animals and plants interact with one another in a great variety of ways, both directly (by eating one another, by overgrowing one another, by dislodging one another, by exuding metabolic products that are poisonous to one another, etc.) and indirectly (by consuming one another's food resources, by dominating the substrate to the exclusion of one another, etc.). The vast majority of all these interactions can be thought of as the products of the factors of competition and predation. The former occurs when different organisms utilize the same environmental commodity (such as competition for food or space); the latter is simply one organism eating another. Just as habitat types and temperature were the principal **physical** forces regulating the distributions of animals and plants, so competition and predation are the major **biological** forces regulating their distributions. Almost all competition may be thought of as competition for food or space. Theoretically, the less competition the members of a species must engage in, the more of their time and energy may be spent on matters other than moment-

to-moment survival, such as growth, reproduction, parental care of young, etc. Further, severe competition between two species with similar environmental requirements may result in the eventual exclusion of one or the other from a particular habitat. Bearing these things in mind, one realizes that a successful organism may not be one that is a "good" competitor, but rather one that is "good at" **avoiding** competition.

There are a number of ways of avoiding competition with closely-related organisms (which are the most likely to be potential competitors), but by-and-large these may all be considered as "specializations" — the more **specialized** the organism, the better it is at avoiding the hassles of competition. The periwinkles which are adapted to life in the spray zone (a specialization) are not subjected to much competition with similar organisms. Similarly, the red alga *Smithora* is specialized to live on the blades of eelgrass and thus has reduced or eliminated competition with most other algae. In both of these examples, the organisms in question avoid competition by living in (having become specialized for) specific habitats in which few other similar creatures reside. The range of specialization of marine life is immense, and obviously some animals and plants are more high-evolved specialists than others. The one drawback to specialization is that a species becomes increasingly more limited in respect to where it can live, and hence it may be expected to occur in lower numbers than the generalists (and be less capable of surviving a major perturbation of the environment). These comments are generalizations — there are exceptions to these "rules of thumb," just as there are exceptions to every "rule" of biology. It is the exceptions that make the study of the natural sciences the most fascinating of all scientific endeavors, for through an understanding of an exceptional situation, a deeper understanding of the "rule" or underlying principle is gained.

The phenomena discussed above are all easily observed in the littoral environment. Take, for example, the case of the scale worm *Arctonoe vittata* (fig. 36), whose entire adult life is spent in the mantle cavities (gill chambers) of certain mollusks and the arm grooves of certain sea stars. *Arctonoe* thus lives a life free from competition with any other similar life-form (and probably quite free from predation also), since no other similar creature resides with it in its special habitat.

Competition does exist, certainly, and a number of experiments and field observations have revealed that it may play a major role in the distribution of marine life on the seashore. There are certain species of barnacles, for example, that are restricted to particular tidal levels by direct competition with other, similar species of barnacles; and the same holds true for certain species of intertidal snails and plants as well. In fact, as mentioned earlier, a number of intertidal organisms have been shown to have their upper tidal limits regulated by physical factors (exposure to air and its consequences), while their lower tidal limit is regulated by competition or predation. Our previously mentioned example of *Pisaster* feeding upon the mussel *Mytilus* is a good example of how predation regulates the lower limit of a prey species. That is, the reason mussels do not occur lower on the rocks is because of more intense predation there by *Pisaster.*

Predation is clearly a powerful force in affecting community composition in the seas. We have already seen how it may regulate species diversity, as well as affecting the natural distribution of a species. It may also play a vital role in more large-scale **geographic** distributions of animals and plants, perhaps in conjunction with competition. Recall that as one moves south along our seacoast new biogeographic regions are encountered, each having in turn more species of animals and plants than the adjacent northern region. This increased abundance of animal types means that there are more (and different) predators and competitors in the southern regions than in the northern regions. An animal that is capable of living in warmer waters to the south of its present distribution may not be able to extend its range simply because it is there confronted with a vast new array of competitors and predators that it is not capable of living with successfully. One example of this phenomenon may be a particular group of isopod crustaceans belonging to the suborder Valvifera. These particular isopods are somewhat specialized and live in intimate association with certain brown algae in the littoral and shallow subtidal waters of California. As one moves south from the Mexican border new but similar algal species appear, and the species of valviferan isopods that are found living on them are quite different; the temperate species are replaced by those of the subtropics and tropics. There is growing evidence being accumulated by the southern author and others that it is not the warmer waters that are keeping the cool-water isopods out of Mexico, but rather some biological factor. This factor is yet to be determined, but it is most likely either competition with the tropical species of isopods, or predation by newly-encountered tropical predators. It seems reasonable to hypothesize that the cool-water isopods, by moving south into Mexico, might be living at or near the upper limit of their range of thermal tolerance, and hence weakened and less capable of competing with the local southern species or of avoiding being eaten by the southern predators. This hypothesis could be tested, in part, by removing all of the warm-water forms from a particular place and introducing a colony of the California isopods to see if they were then able to form a successful, reproducing population in the absence of local competition.

When you visit the seashore, take the time to sit quietly on the edge of a tidepool and just watch. Within 5 or 10 minutes you will be able to observe predation in action. If you are a patient and careful observer, you will be able to record in your field notebook a variety of escape responses of the various prey animals. If you are thoughtful, you may also be able to make judgements as to how specialized, or generalized, various species of animals might be, and whether or not they may be actively competing with others, or avoiding competition.

ARTIFICIAL FACTORS

Our coastal environment is suffering increasing damage from the effects of a great variety of man-made alterations. These include coastal construction of all kinds which change or destroy shore habitats (marinas, harbors, housing and commercial developments, docks, piers, landfills, etc.), as well as other kinds of ecological alterations which may also influence the animals and plants of the nearshore waters, either directly or often indirectly in more subtle ways. The latter category includes the activities and effluents of such things as sewage-treatment plants, areas of contaminated rain or river runoff, petroleum contaminants, chemical and power plants, pulp mills, etc., which may cause various degrees of pollution of different kinds. In general, all of these pollutants tend to alter the environment and typically reduce the number of, or change the kinds of organisms which live there. Some naturalists tend to consider any alteration of the natural environment as a sin against the Earth itself. Clearly, reason must be incorporated into this kind of thinking, for a modern society must provide itself with the space to live and to produce the essential materials upon which it subsists. The authors, therefore, do not subscribe to the extremist or radical approach to "ecology" as a fad, but rather believe that ecological studies of high quality should serve as sources of information to guide the necessary change and reduce the risk of environmental damage. However, there are certain natural environments that are very limited, often maintaining a tenuous existence balanced between two much larger and different environments. The seashore is such an area, being that narrow strip of habitats lying between the open sea and land.

This thin ribbon of rocks and sand and surf is unique in that it supports a greater species abundance than any other life zone of similar size on Earth. The littoral and shallow sublittoral regions also provide refuge for the animals and plants that are responsible for the nearshore productivity of the world's oceans, which in turn support the important commercial fisheries of the world. But the littoral region has additional, more far-reaching significance. The coastal bays and lagoons serve as essential spawning grounds for many marine fishes and shellfishes, and as nurseries for the young of these and other animals. These same semi-enclosed bodies of water are frequented by various water fowl each year during their spring and fall journeys to their feeding and nesting grounds. Estuaries serve as nutrient traps, sinks into which enormous amounts of vital nitrates, phosphates, and other substances are stored and released slowly into the sea. The list of the seashore's critical roles goes on and on, but we cannot omit mention of its educational and recreational value. Every year millions of Americans visit the coastal environment, for a multitude of reasons: fishing, swimming, surfing, SCUBA diving, solitude, escape from the cities, and often for youngsters as a first real introduction to the natural world. It is a place where we have, time and again, seen the natural inquisitiveness of youth blossom with the joy of discovery. The seashore is, without a doubt, nature's greatest outdoor school, open to all at no cost, save the time it takes to visit the nearest tidepool or beach.

So, it would seem that the only logical approach to the use of this unique and important gift of nature would be to use the seashore as necessary to provide society with its basic common needs, but to do so in a way which alters these fragile environments as little as possible, and to set aside sufficient area as undisturbed to insure the shore's preservation. Unfortunately, such simple reasoning rarely prevails. Those parties that have traditionally been responsible for coastal alterations have done so with a single motive — quick profits. Let it be known that the authors are by no means anti-profit, but there is certainly a multitude of paths to monetary gain other than those which destroy the beauty and long-term value of coastal environments for the sake of a fast buck.

The Earth is presently experiencing the first waves of what some have termed a worldwide ecosystem collapse. A mass devastation of animal and plant species is being brought about almost solely by the efforts of man, largely through the destruction of natural environments. Literally tens of thousands of life forms will become extinct on this planet in the next 50-100 years, if present trends of population growth, urban expansion, and resource exploitation continue. The areas of the world now experiencing the most rapid rates of growth are those countries located in the tropical and subtropical regions. Unfortunately, these are the same regions that are biologically most poorly understood, and we can expect increasing rates of extinction of animals and plants, some of which have never even been named or described, and many more whose role in the balance of nature has yet to be determined. Man is not a separate biological entity on this small satellite of the Sun, but he is without a doubt the most stubborn and destructive of all life forms. It would seem that only as these trends of growth and over-exploitation begin to steal our air and food, discolor our land and water, and finally disfigure our children will we awaken to the real oneness of life and our place in the natural scheme of things. Even that may be an optimistic prediction.

Today's worldwide phenomenon of coastal urbanization is probably more threatening to natural communities than any other of man's activities, with the possible exception of uncontrolled marine pollution. By way of example of the former source of destruction, Tampa Bay, Florida, has already lost about 25% of its original water area to land fill, while San Francisco Bay, California, has lost about 35%. These losses result in reduced spawning areas for fishes and invertebrates, reduced overwintering regions for migrating water birds, changes in local weather conditions, and long-term reductions in local offshore productivity. Virtually no estuaries or coastal lagoons remain unaltered in California, and all but a very few have been subjected to serious damage.

The changes wrought by man upon our seacoasts invariably create new and different kinds of local environments, where only certain select kinds of animals and plants can exist. Docks and piers are good examples of artificial habitats, and a few minutes spent examining the community that inhabits a floating dock will reveal an assemblage of organisms rarely found elsewhere in the intertidal. This assemblage has been called the **fouling community**, and it may also be found on ships' bottoms and other similar submerged substrates. Such a community contains all of the necessary components for the full operation of a biological system: producers (various algae), grazers and filter feeders (gastropods, mussels, etc.), high-level carnivores (seastars), and a host of smaller creatures.

CHAPTER IV

IDENTIFICATION OF INTERTIDAL ANIMALS

Of the 25 or 30 major animal groupings or **phyla** (singular **phylum**) recognized, all but one are composed exclusively of invertebrates, those creatures which do not possess a backbone. The single exception (Phylum Chordata) includes the more familiar backboned types (fishes, amphibians, reptiles, birds, and mammals) within its subphylum Vertebrata. The system of animal and plant classification is made up of a hierarchy of categories from the Kingdom (Animalia or Plantae) down to the level of species, at which organisms are precisely identified. Each category is called a **taxon** (plural **taxa**). The structure of these classifications incorporates our knowledge of the evolutionary relationships of the groups in question. To illustrate this system, the California mussel, razor clam, moon snail, and giant green anemone are classified below (Table 1). The degree of evolutionary relationship between these groups is reflected in the number of shared taxa. For instance, the mussel and razor clam share the first three taxa, and are thus more closely related to each other than either is to the moon snail, with which they share only two taxa. All three of those animals share only one taxon with the anemone.

Notice that the **generic name** or **genus** (plural **genera**) is capitalized and placed in italics (or underlined if italic type is unavailable). The **specific name (species** is both plural and singular) consists of the generic name and a non-capitalized **specific epithet** or **trivial name,** both of which are italicized. This double naming of organisms is called the System of Binomial Nomenclature and is governed by a great number of rules and regulations. You will see, in more detailed books, the specific names of organisms followed by the name(s) of the individual(s) who first described the particular species, and dates indicating when the original naming took place. Long sessions are conducted, many pages printed, and numerous feathers ruffled over whose name has priority in cases of redescription or revisions of a classification; the animals and plants, however, do not care about these legal battles at all.

We feel obliged at this point to say something about the use of common or vernacular names. It is not ivory-tower isolationism which forces professionals generally to avoid more colorful common names, which are easily pronounced and spelled, in favor of scientific names for animals and plants. Rather it is the inevitable confusion which results from regionally different common names for the same creature (e.g. horseneck and gaper clam for *Tresus*), or a single such term which applies to more than one species. In some cases a single common name (e.g. clam) may include a multitude of species. While a few groups of animals, notably the birds, have reasonably solid

Table 1: Classifications of the California mussel, razor clam, moon snail, and green anemone.

	mussel	razor clam
Kingdom:	Animalia	Animalia
Phylum:	Mollusca	Mollusca
Class:	Bivalvia	Bivalvia
Subclass:	Pteriomorpha	Heterodonta
Order:	Mytiloidea	Veneroida
Family:	Mytilidae	Solenidae
Genus:	*Mytilus*	*Siliqua*
Species:	*Mytilus californianus*	*Siliqua patula*

	moon snail	green anemone
Kingdom:	Animalia	Animalia
Phylum:	Mollusca	Cnidaria
Class:	Gastropoda	Anthozoa
Subclass:	Prosobranchia	Zoantharia
Order:	Mesogastropoda	Actinaria
Family:	Natacidae	Actiniidae
Genus:	*Polinices*	*Anthopleura*
Species:	*Polinices lewisii*	*Anthopleura xanthogrammica*

vernacular naming systems, most do not. Scientific names, on the other hand, are standardized and applicable anywhere in the world. In addition, the Latinized terms of the binomial actually tell the reader something about the animal or plant in question (or they should, at least). We regret that Latin is an infrequent inclusion in the class schedules of most students.

Note that there is a level of subclass in the examples given in Table 1; such intermediate categories are common and reflect the size of the group in question and the diversity of its members (i.e. the larger and more complex the group, the more subdivisions are needed to classify its members). At any rate, as one reads from higher to lower in the hierarchy the organisms contained in each successive grouping are more and more alike, until, at the level of species, we are dealing with those creatures which look, behave, and live alike and are capable of natural interbreeding.

The identification of some organisms in the field can occasionally be difficult, especially for the casual observer. After only a few carefully undertaken excursions, however, you should be able to place most animals in their proper phylum or class on sight. It may take a little practice, but we guarantee that your ability to recognize types of organisms will steadily improve. The following section is devoted to the identification of seashore invertebrates to the level of phylum and, in some cases, class. Each group is discussed in more detail, along with specific examples, in a later chapter. Some of the common vertebrates and seashore plants are treated later.

Biologists have devised an artificial tool called a "key" for the identification of organisms. Keys come in all forms and are constructed to deal with various levels of classification. The key below will help you to identify almost any invertebrate to its phylum or class, from which you can proceed to the appropriate pages which treat that group. Our key should work for most of the common animals you encounter along the north coast, but there will be exceptions, so always check your decisions against the figures and descriptions. Notice that the key is in the form of couplets or pairs of choices. There are two choices at number "1," two at number "2," and so on. Read each member of the couplet, beginning at number "1," and choose the one which best fits the animal under question. Following your choice you will find either a name (presumably the correct one and you can proceed to the appropriate pages of descriptions) or a number (in which case you proceed to that numbered pair of choices further along in the key).

The authors are compelled to remind the reader that the size constraints and general nature of this book are such that it has been impossible to include more than the most common types of local seashore organisms. For those readers who are stimulated to pursue further their interests in animals and plants of the Pacific Coast, we strongly urge you to investigate some of the references listed at the back of this book. Some of the best sources to use as a beginning to expand your knowledge are: Abbott and Hollenberg, 1976 (for algae); Hedgpeth, 1962, Kozloff, 1973, and Ricketts and Calvin, 1968 (for invertebrates); and Young, 1949, MacGinitie and MacGinitie, 1949, and Carefoot, 1977 (for general seashore ecology).

KEY TO THE MAJOR GROUPS OF INTERTIDAL INVERTEBRATES

1. Worms and worm-like animals; body thin or sausage-shaped and elongate, or oval and flattened; free-living, burrowing, or in tubes (figs. 32-44, 121, 137-139) 2
1. Not worms or worm-like; body not as described above 8
2. Very flattened oval worms; 2-3 times as long as wide; (1-3cm long, 1-2mm thick); (fig. 32); the flatworms. . . . . Phylum Platyhelminthes, Class Turbellaria
2. Body not flattened and oval, usually round or oval in cross section; generally many times longer than wide 3
3. Body of worm composed of rings or segments separated by grooves; may be free-living, burrowing, or in tubes of various sorts (figs. 35-42) . . . . Phylum Annelida, Class Polychaeta
3. Body of animal not obviously composed of segments or rings . 4
4. Body generally rather thick, sausage-shaped; adhering to hard substrates by tiny suction cups at the ends of numerous flexible tubes (figs. 137-139); the sea cucumbers. . . . . Phylum Echinodermata, Class Holothuroidea
4. Not as described above; body without sucker-like tubes. 5
5. Worms living in long narrow vertical sand tubes (up to 30cm long and 4-5mm diameter) in bay mud and sand flats; small flower-like tuft of green tentacles at upper end of body (fig. 121). . . . . Phylum Phoronida *(Phoronopsis viridis)*
5. Not in thin, long tubes; without green tentacles 6
6. Extremely flexible and extensible worms, many times longer than wide; body rubbery; often with bright or distinctive coloration (figs. 33-34); the ribbon worms. . . . . Phylum Nemertea
6. Body not extremely thin and elongate or rubbery. 7
7. Relatively large (10-25cm long) soft, pink, sausage-shaped worms; burrowing in tidal flats (fig. 44). . . . . Phylum Echiura *(Urechis caupo)*
7. Body smaller, more rigid, not distinctly pink; usually found in mussel beds or burrowing in sand under rocks (fig. 43); the peanut worms . . . . Phylum Sipuncula
8. Firmly glued to the substrate, cannot be removed without damaging the animal (not just attached by suckers or suction apparatus). 9
8. Not actually glued to the substrate 17
9. Body encased between two shells (figs. 110-113, 122) 10
9. Body with 0, 1, or more than two shells 11
10. Animal cemented to substrate by one shell or attached to sub-

strate by numerous tough threads (figs. 110-113); mussels, scallops, oysters. . . . . some members of the Phylum Mollusca, Class Bivalvia

10. Animal attached to the substrate by a fleshy stalk arising near the hinge of the shells (fig. 122); the lamp shells . . . . Phylum Brachiopoda

11. Animal encased in several white to gray calcareous shells or plates; either volcano-shaped and attached directly to the substrate, or with a tough leathery stalk (figs. 47-50); the barnacles . . . . . Phylum Arthropoda, Class Crustacea, Subclass Cirripedia

11. Not as described above. 12

12. Body more or less cylindrical in shape; attached to substrate by a flattened base; generally fleshy with a ring of tentacles at the upper end (figs. 25-30); the sea anemones . . . . Phylum Cnidaria, Class Anthozoa

12. Not as described above; body either bushy or feathery, usually drab, pale brownish or grey, or forming a thick or thin encrustation on solid substrate, or growing singly or in clumps of finger-like individuals and lacking tentacles (figs. 10-11, 15-18, 123-125, 140-142). 13

13. Surface slick or slimy to the touch; often brightly colored (figs. 141-142); the social and compound ascidians . . . . Phylum Chordata, subphylum Urochordata

13. Surface not slick and slimy to the touch. 14

14. Thin to thick mats or encrustations on solid substrates, or clumps of finger-like projections; rough and spongy to the touch; surface usually bears obvious holes (figs. 10-11); the sponges. . . . . Phylum Porifera

14. Not as described above, if mat-like then very hard and calcareous 15

15. Very thin, hard encrustations on solid substrates; often red or yellowish in color; surface with tiny net-like pattern (fig. 124); . . . . some members of the Phylum Ectoprocta (=Bryozoa)

15. Erect (not mat-like or encrusting); growing either singly or in bushy or feathery bodies (figs. 15-18, 123, 140). 16

16. Growing singly as an attached leathery rounded or elongated body; with or without a thinner stalk between the main body and the substrate; usually taller than 3cm; with two openings at the upper end (fig. 140); the solitary ascidians. . . . . Phylum Chordata, subphylum Urochordata

16. Small, usually delicate, bushy or feathery bodies (figs. 15-18, 123); the hydroids (Phylum Cnidaria, Class Hydrozoa) and the

arborescent bryozoans (Phylum Ectoprocta) (See later sections and figures for distinctions between these two groups.)

17. Body encased in a rigid but jointed outer skeleton; with jointed legs (figs. 45-46, 51-71); crabs, shrimps, and various bug-like creatures .... Phylum Arthropoda
17. Body without obviously jointed appendages 18
18. Body jelly-like; usually washed ashore on beaches or tidal flats. 19
18. Body not jelly-like 20
19. Disc- or umbrella-shaped (figs. 20-23); the jellyfishes .... Phylum Cnidaria
19. Body appears like a small (1cm diameter) clear jelly-like marble (fig. 31); the comb jellies and sea walnuts .... Phylum Ctenophora
20. Body disc-shaped, star-shaped, or globular to nearly spherical; often spiny (figs. 126-136); the seastars, brittle stars, sea urchins and sand dollars .... Phylum Echinodermata ... 21
20. Body not as above 22
21. Body star-shaped with five or more arms radiating from a central disc (figs. 126-133, 136) Classes Asteroidea and Ophiuroidea
21. Body globular or disc-shaped (figs. 134-135)... Class Echinoidea
22. Slug-like, or snails with a single cap-shaped or coiled shell (figs. 79-109); sea snails and sea slugs .... Phylum Mollusca, Class Gastropoda
22. With or without shells, but not slug-like or single-shelled 23
23. With eight shell plates along back, and a large snail-like foot on underside (if no plates visible then back warty, reddish brown, up to 30cm long — *Cryptochiton stelleri*, fig. 72) (figs. 73-78); the chitons or sea cradles .... Phylum Mollusca, Class Polyplacophora
23. Without eight shell plates along back 24
24. Soft-bodied; with eight suckered arms; octopuses (squids have ten arms and are not intertidal) .... Phylum Mollusca, Class Cephalopoda
24. Body encased in two shells (figs. 114-120); clams .... Phylum Mollusca, Class Bivalvia

CHAPTER V

ANIMAL STRUCTURE

A crusty old Botany professor with whom we were once acquainted used to begin his classes with this bit of philosophy. "Before studying this subject you view a tree and it is, in your minds, just that — a tree. During this course that object will become a collection of parts and processes, cells, tissues, and energy flow, until it eventually becomes a tree again. The tree you know now and the one you will know at the end of this class, however, are two very different things." All organisms, including the creatures of the seashore, can be better understood and more appreciated with an increased knowledge of how they are put together and how they work.

Every animal, no matter how simple, must accomplish a number of fundamental feats to survive. Their basic quest is to stay alive long enough and in a healthy enough condition to produce offspring. In order to do this they must eat, exchange gases with their environment, excrete waste products, etc. Once they have passed their age of reproductive ability they have, biologically speaking, nothing else to contribute to their species, and they die. Modern man is an exception to this principle, and we now manage to keep each other alive long after we are finished replacing ourselves with offspring. The strategies employed by animals to survive are as varied as the creatures themselves, and while this book is not intended to be a text in Zoology, some basic discussion of animal body types is appropriate and should lead to a better understanding of how the animals you find are equipped to handle the problems of survival. In addition, our knowledge of the evolutionary relationships between various animal groups is based largely on how the beasts are put together and how those structural forms develop during the embryonic stages. Our classification systems are, in turn, based upon these very real relationships, although we still lack knowledge concerning many details of the evolution of most groups. These evolutionary and classificatory schemes are continually being changed as new information becomes available, or as old information is interpreted in different ways. At any rate, anatomy is anatomy, and throughout the animal kingdom there are some basic trends which are worth knowing about.

SYMMETRY

The regular arrangement of an animal's body parts in relation to the axis of its body is called **symmetry**, and several types are easily recognized. Some animals are **asymmetrical** (many sponges and protozoans) and show no regularity of form about an axis at all. Others have their body parts arranged radially around the body axis and thus display **radial symmetry** (jellyfishes, fig. 22) where the body may be sliced along its axis on any plane to produce similar halves. There are several types of radial symmetry. One type is exemplified by a five-armed seastar which shows **pentaradial symmetry** (having the body parts arranged along five radii around the axis, fig. 128). Other animals, such as many sea anemones, display **biradial symmetry** (fig. 27), where the body may be sliced along its axis on only two planes to produce similar halves.

Most familiar animals have **bilateral symmetry**, there being only a single plane of sectioning which separates the body into similar halves. Bilateral animals, unlike radial ones, generally have a **front** end (**anterior**) and a back end (**posterior**), and right and left sides to the body. Further, if bilaterally symmetrical animals move around, they tend to do so in one direction, with the anterior end forward. As obvious as this condition may seem, it has some far-reaching and important implications. In order for such an animal to survive, all of its systems must be structurally and functionally coordinated with its symmetry and unidirectional movement. For example, the nervous system becomes concentrated as some sort of brain, along with the major sensory organs, at the anterior end to form a head (a phenomenon known as **cephalization**). Thus the animal gathers information about its environment as its anterior end enters new conditions. It would certainly not be advantageous to be constructed in such a way that the body was unaware of adverse conditions until the tail came along to find out about them! So the bilateral animals are cephalized to varying degrees, while the radially symmetrical ones have no "brain" as such, and have their sense organs dispersed all around their bodies. Bilateral symmetry is adaptive for animals which move about, as described above, whereas radial symmetry is generally adaptive for more sedentary or drifting animals, by allowing them to receive information from any direction. Similar correlations can be drawn between symmetry and other features such as feeding methods, gut structure, muscle arrangement, etc.

CELLS, TISSUES, AND GERM LAYERS

The animal kingdom may be divided into those animals which have their bodies organized as and developed from integrated layers of cells called tissues (the **Metazoa**), and those which do not (the **Protozoa**). The latter includes only the single-celled creatures such as amebae, and a few which form aggregations of cells of similar structure and function. The Metazoa include all other forms of animal life, with the possible exception of the sponges (Phylum Porifera). Many zoologists place the sponges in a third category, the **Parazoa,** suggesting for a variety of reasons that these creatures do not possess true tissues. Their embryogeny is not easily compared to those patterns seen in any metazoan, especially as regards the formation of tissues. Also, sponge cells display uncanny abilities to become de-specialized and then form other types of needed cells, or entire new sponges, their fates never becoming completely fixed unlike those of true metazoans whose cells have little or no ability to change from one type to another.

Having conveniently dispensed with the protozoans and the sponges, we can now proceed to a discussion of the arrangement of tissues within the true metazoans. The key to understanding the levels of complexity within the Metazoa is found in their early embryogeny. All true metazoans pass through some comparable developmental stages when the first layers of tissues arise. These embryonic tissues are called **germ layers**, and from them all adult structures form. Some groups, the Cnidaria and Ctenophora (jellyfishes, sea anemones, comb jellies, and their kin) form and forever possess the derivatives of only two germ layers, called the **ectoderm** (outer layer) and **entoderm** (inner layer). Such animals are termed **diploblastic** or two-layered. All of the other major animal phyla develop a third germ layer called **mesoderm** (middle layer), which lies between the other two tissues and defines the animals as **triploblastic** (three-layered). Generally speaking, the ectoderm forms the outer skin and the nervous system; the entoderm forms the digestive tube and its associated structures; and the mesoderm forms such things as the circulatory system, complex musculature, and parts of the reproductive system.

MESODERM AND THE BODY CAVITY

There are different levels of complexity and body organization within the triploblastic animals, depending primarily upon the fate and arrangement of the mesoderm and the presence or absence and nature of a cavity (**coelom**) between the inner and outer body layers. The flatworms and ribbon worms (Platyhelminthes and Nemertea) have a more or less solid layer of mesoderm between the gut and the outer body wall, and are hence referred to as being **acoelomate** (without a body cavity, fig. 5). Seven phyla which include such things as rotifers, roundworms, and other odd groups of small beasties do have a body cavity between the ectoderm and entoderm, but it is incompletely lined with mesodermal tissue, and thus referred to as a false body cavity, or **pseudocoelom** (fig. 6). All of the other triploblastic phyla may be considered **eucoelomate** (or simply **coelomate**) in that they possess a "true body cavity" filled with fluids and fully lined by a thin layer of mesodermally-derived tissue (fig. 7). This membranous lining is usually referred to as the **peritoneum.** To confuse the issue somewhat, it should be noted that some of the more advanced phyla within this latter category (Arthropoda and

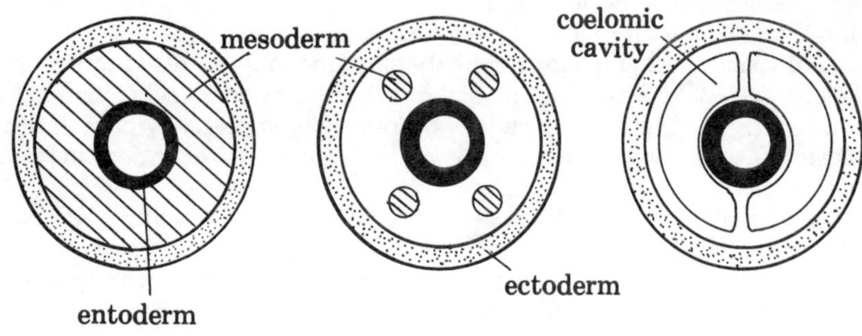

5. acoelomate 6. pseudocoelomate 7. coelomate

(all diagrammatic)

Mollusca) do not display a well-developed coelom as adults, but a great deal of evidence has convinced us that it was there in their ancestors and is still present in modified form. The separation of the inner and outer layers of the body by fluid-filled spaces has allowed the realization of a variety of advantages and increases in complexity when coupled with certain body forms. For example, the protrusion of certain body parts, such as the feeding proboscis of some worms, is accomplished by increasing the hydrostatic pressure within the coelomic fluid. Particular types of crawling and swimming are only possible through the use of the coelomic spaces in conjunction with complex musculature.

In addition to the above differences, animal groups are variable in less basic but perhaps more obvious ways. Some animals have both a mouth and an anus (a **complete gut**), while others have but a single opening for things entering and leaving the digestive system (an **incomplete gut**); some have a well-developed circulatory system, others have none at all; and so it goes. Some of these features are pointed out in the general descriptions of each phylum. All of the foregoing may have little to do with finding or naming the clam or snail you find at the seashore, but it does have a great deal to do with how these animals go about accomplishing their fundamental feats of survival, by being anatomically equipped for and adapted to particular habitats and ways of life, and we are thus obliged to include it. As Jerome Tichenor has told us:

> "If this leaves the student all at sea,
> It's not because the teacher has not tried:
> In his lectures, as the old prof said to me,
> The Metazoa are made of layers three."
>
> (From **The Party Line.** In: Poems In Contempt of Progress. Boxwood Press, 1974)

CHAPTER VI

THE INVERTEBRATES

The Greek or Latin derivations are given after each phylum name.

PHYLUM PROTOZOA (first animals)

the single celled animals

All of the major groups of protozoans are well represented in the sea, but only a few of the intertidal forms are large enough to be seen even with the aid of a hand lens. The **foraminiferans** (a group of shelled amebae) are extremely abundant and important members of the plankton but are also represented on the north coast by two relatively large intertidal forms. *Gromia oviformis* may be seen with the naked eye as tiny (2-3mm) brown spheres adhering to the undersides of rocks, blades and holdfasts of algae, and to other solid surfaces in protected areas. Thin extensions of the cell, called **pseudopodia,** can be extended through openings in the surface of the sphere and are used both for movement and to capture tiny bits of organic material on which the protozoan feeds. A second species, *Discorbis patelliformis,* may be identified as a small (1mm) white calcareous shell. The coiling of the shell gives this animal the appearance of a tiny snail. Both of these protozoans are eaten by a variety of small creatures which browse over the surface of rocks and algae.

A number of other protozoans visit our beaches from near-shore waters during various periods of the year. Included here are members of a group known as the **dinoflagellates** (these organisms are capable of photosynthesis and thus also enjoy being classified within the plant kingdom in the Division Pyrrophyta). If you have a microscope at your disposal it will be possible for you to observe the creatures responsible for an exciting natural phenomenon by examining water samples from along the shore. Watch for changes in water color; patches of reddish-brown water often indicate high concentrations of dinoflagellates. Of local importance is *Gonyaulax* (fig. 8), the culprit involved in producing "red tide" and its associated

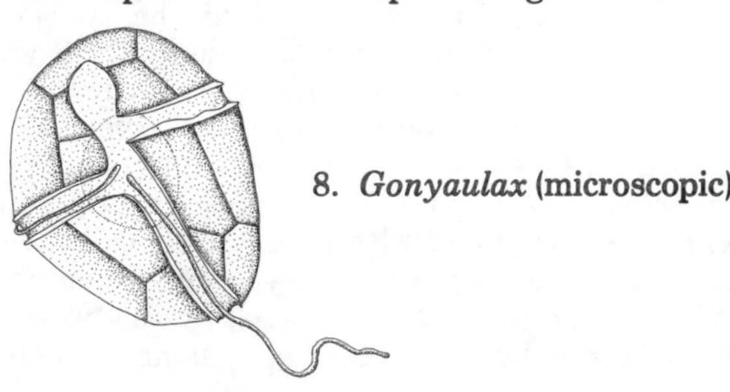

8. *Gonyaulax* (microscopic)

mussel or shellfish poisoning. The toxic substance produced by this protozoan is concentrated in mussels and numerous other clams which feed by filtering the sea water. Although the mollusks are not harmed by consuming *Gonyaulax,* any vertebrate, including man, which eats them may suffer serious effects. The major increases in concentrations of *Gonyaulax* (referred to as "**blooms**") occur during the early and mid-summer months when conditions for growth and reproduction are optimal. However, along northern California blooms may not be very regular or predictable, and localized short-term blooms may be caused by excess nutrient runoff from rivers or by other contributing factors. It is always a good idea to check with the County Health Department or other local Public Health Office if you plan to prepare shellfish for a meal, and be sure that a quarantine is not in effect.

PHYLUM PORIFERA **(pore-bearing)**
the sponges

In the majority of cases the specific identification of sponges is best left to an expert in such matters, although with these animals even the experts do not always agree upon who's who. To become accomplished in recognizing species of sponges, one must learn the language of sponge specialists and master some laboratory techniques to remove and identify the microscopic skeletal elements (**spicules**) of the creatures, all of which takes more time than most of us are willing to spend. The inability to name a sponge, however, should not detract from one's enjoyment and appreciation of the exciting and unusual nature of these animals.

Sponges are generally found as clumps or thick sheets attached to hard substrates, expecially large rocks. These mats of sponges are generally somewhat rough or crusty to the touch, rather than slick or slimy. A few simple sponges exhibit radial symmetry, but most grow irregularly in all directions and tend to conform to the general shape of the surface on which they live. "Colonial life" is the rule among sponges, and it is difficult to tell where one individual begins and another leaves off. The colony may consist of hundreds of individuals which are connected within the mass of the sponge by complex canal systems. A few species retain their individuality and generally take the form of finger-like tubes arising as a cluster from a common base, or occasionally, of a solitary vase-shaped form.

A sponge colony is full of holes, some large and easily seen, many more small and microscopic. These pores lead into a central chamber or series of chambers which are lined with unique cells called **collar cells** or **choanocytes** (fig. 9), each with a long whip-like extension called a **flagellum**. Water is drawn into the sponge, by the beating of these flagella, through the smaller holes called **ostia** (singular **ostium**)

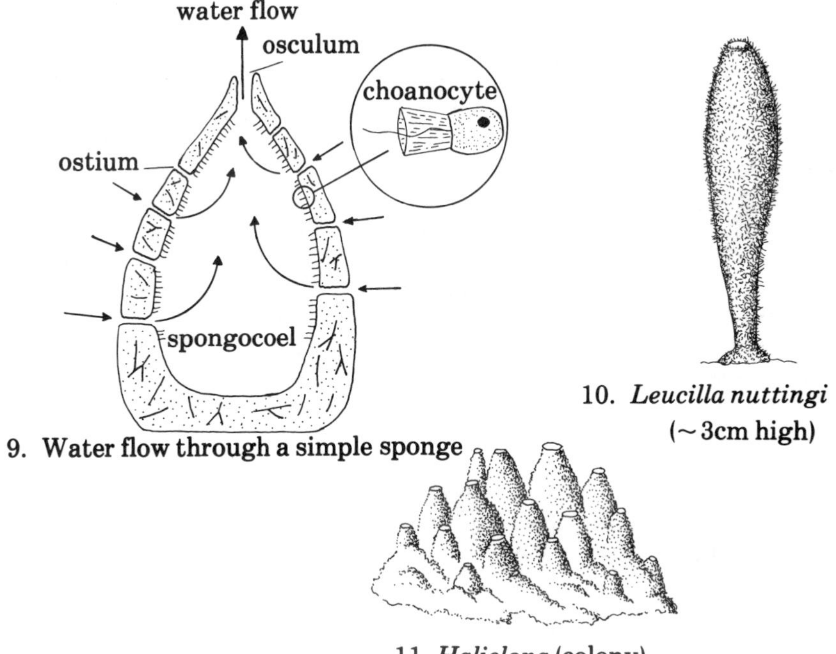

9. Water flow through a simple sponge

10. *Leucilla nuttingi* (~3cm high)

11. *Haliclona* (colony)

and forced out through the larger openings called **oscula** (singular **osculum**). The choanocytes not only produce the water current through the sponge, but also, as the water passes over them, extract food such as minute particles of organic detritus and, to a lesser extent, bacteria and other single-celled organisms. The food particles are taken into the choanocyte and then passed to motile cells which transport the food throughout the body. Waste products are released into the outgoing water that leaves the sponge via the oscula. Sponges which live in calm water often have the oscula raised on volcano-like extensions so that the waste water is less likely to be recycled through the body. In turbulent situations the waste is rapidly swept away, and sponges living in such conditions have the oscula nearly level with the main body of the colony.

Relatively few animals feed on sponges, probably because of their internal spicules and perhaps bad taste resulting from the secretion of noxious substances in some species. Certain gastropods (snails and sea slugs) and a few species of fishes graze on sponges. Possibly the most serious danger facing an adult sponge is having its filtering system clogged with sediments. For this reason, and because they cannot attach to soft substrates, sponges are rarely found in muddy habitats.

Sponges reproduce both asexually (by budding leading to colony growth) and sexually. In the latter case, sperm are released into the sea water from which they are picked up in the water currents of neighboring sponges. Once inside the sponge the sperm are taken into the collar cells (as in feeding). Rather than digesting the sperm, however, the collar cells leave their position in the lining of the inside of the sponge and carry the sperm to eggs which are imbedded within the sponge's body. Following fertilization and early development, a larva is produced which leaves the body of the parent, swims about for a short time, and eventually settles and metamorphoses to a young sponge.

The skeleton of sponges is the key to species-level taxonomy, but it is of little or no use in field identification. The various shapes of these needle-like spicules can be seen by smearing a piece of sponge on a microscope slide and treating it with a few drops of bleach before observation. The bleach dissolves away the cells, leaving the spicules easily visible. Many of the very thick sponges have a fibrous material (**spongin**) making up most or all of their skeleton along with or instead of the spicules. A true (not synthetic) commerical bath sponge is actually the spongin remains of a sponge after the cellular material has been removed. These commercial sponges are called the keratose sponges (members of the subclass Keratosa of the Class Demospongiae); they are poorly represented in the eastern Pacific. The scarcity of keratose sponges on the world market several decades ago stimulated the manufacture of artificial sponges of cellulose and plastics. Thus, marine biologists may owe a tip of the hat to certain corporations for the salvation of the Carribbean and Mediterrranean "commercial" sponge fauna.

While the amateur or semiprofessional naturalist cannot expect to be able to analyze spicules and thus name all of the sponges he or she encounters, there are two which can be recognized easily by sight and serve as examples of distinct body forms. The only non-encrusting sponge likely to be encountered along the north coast is *Leucilla nuttingi* (once known as *Rhabdodermella*), shown in Figure 10. This sponge (belonging to the Class Calcarea) forms clumps of whitish vase-shaped individuals, two to three centimeters in height, each topped with a large and obvious osculum. Look for *Leucilla* in potected areas such as on pilings and floats, or under ledges in the low intertidal zones, out of the direct force of the waves. The genus *Haliclona* (fig. 11) is represented locally by sponges which form lavender-colored sheets across rock surfaces and other hard substrates. This sponge belongs to the Class Demospongiae, as do most intertidal sponges. There are numerous other common genera along our shores, such as *Halichondria* which forms thick yellowish masses on floats and pilings. Of particular interest is *Cliona*; the members of this genus bore into mollusk shells, especially those of abalones (*Haliotis*) and rock scallops (*Hinnites*) (figs. 79, 113). The shells of these mollusks are often riddled with small holes in which the bright yellow sponge can be seen.

PHYLUM CNIDARIA (with cnidocytes = nettle-cells)

also known as Coelenterata; the sea anemones, jellyfishes, corals, etc.

There are three classes within the phylum Cnidaria, all of which are well represented along the northern California coastline. The Class Hydrozoa includes the hydroids and the hydromedusae, the Class Scyphozoa contains the true jellyfishes, and the Class Anthozoa includes the sea anemones and corals. In spite of the great diversity of form and habits among the cnidarians they are all built on the same body plan. These diploblastic, basically radially symmetrical animals consist of a double-walled sac with the inner (derived from entoderm) and the outer (derived from ectoderm) layers separated by a layer of various thicknesses (fig. 12). This middle layer is composed of a gel-like substance (**mesoglea**) which usually contains various cells. When cells are present the middle layer is referred to as a **mesenchyme**, but because of its embryological origin, it is not considered to be true mesoderm.

Cnidarians and other radially symmetrical animals are constructed in such a way that the axis of the body runs from the side of the body which bears the mouth (the oral side) to the side opposite the mouth (the aboral side), and are thus said to be oriented on an oral-aboral axis. The oral area bears the mouth which is surrounded by a ring of tentacles. These tentacles, and often other body regions, contain unique cells called **cnidocytes** (from which the phylum name is derived), which in turn contain structures known as **nematocysts** (fig. 13). The microscopic nematocysts can be fired into the environment by everting the thread-like portion as shown, and they function in such activities as attachment, prey capture, and defense. Many types of penetrating nematocysts contain a nerve toxin which causes paralysis and these are used for stunning prey. The stings of some

12. Generalized cnidarian body

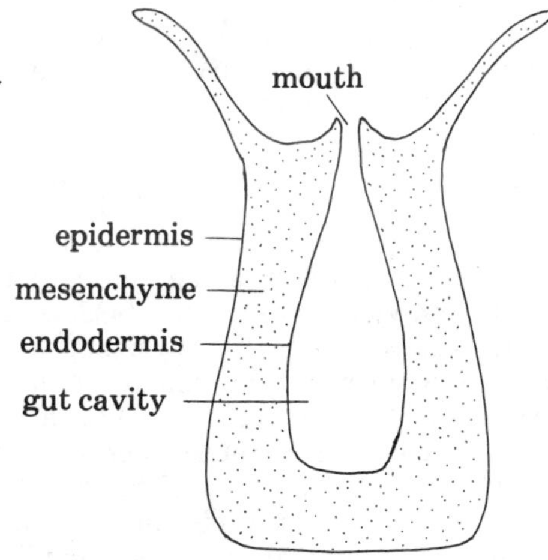

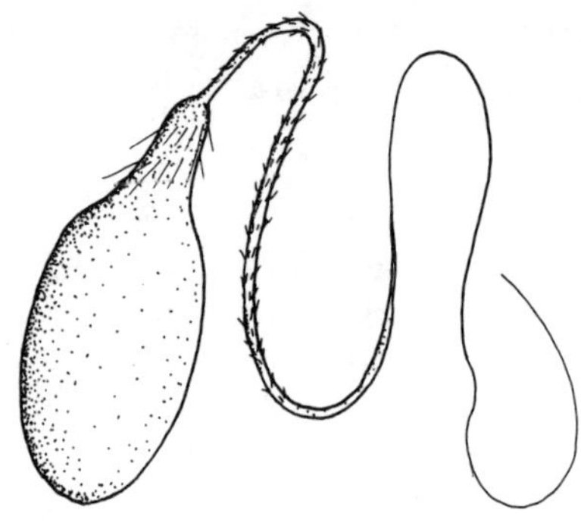

13. nematocyst (microscopic)

cnidarians (especially certain jellyfishes and corals) may cause serious reactions, even death, in humans.

Cnidarians are nearly all carnivorous. Stunned prey animals are taken into the digestive cavity through the mouth, which also serves as an exit for undigested materials. Some of the colonial cnidarians, expecially corals, are filter feeders and create a water current in and out of the mouth from which food material is extracted. A number of cnidarians have certain dinoflagellates (called zooxanthellae) and green colored algae (called zoochlorellae) living within their tissues. Such close associations are termed **symbiotic** relationships, a term which also includes such situations as parasitism. The brownish or greenish colors of some anemones and corals are due to the presence of these microscopic symbionts. In fact, the presence of these unicellular partners is actually necessary for certain aspects of the formation of coral reefs. The cnidarians benefit by receiving oxygen and certain organic compounds from the symbiotic protozoans, which in turn receive carbon dioxide and various nutrients from the metabolism of their hosts.

Intertidal cnidarians are preyed upon by few other animals. Those which do feed on them include some sea slugs (figs. 103-105) and certain sea spiders (fig. 45).

Most hydrozoans and scyphozoans pass through two very distinct stages in their life cycles (fig. 14). In general, the members of these two classes are at one time sessile, attached **polyps**, which bud off small jellyfishes (called **medusae**) by asexual means. The medusa stages then produce eggs and sperm which unite (sexual reproduction) and grow into characteristic ovoid larvae called **planulae**, which eventually settle to become yet another polyp. The free-swimming medusoid or jellyfish stage has been abandoned by the anthozoans, and the polyps can reproduce both sexually and asexually.

In keeping with the style set by radial symmetry, cnidarians possess a diffuse, non-centralized nervous system. Many of these creatures, however, are capable of rather complex behavior in spite of their simple arrangement of nerves and sense organs. Sea anemones, for example, display very different reactions to edible and non-edible objects placed upon their tentacles (try it!). We cannot always explain how an animal manages complicated activities with relatively simple structural equipment, but it does give zoologists something to think about.

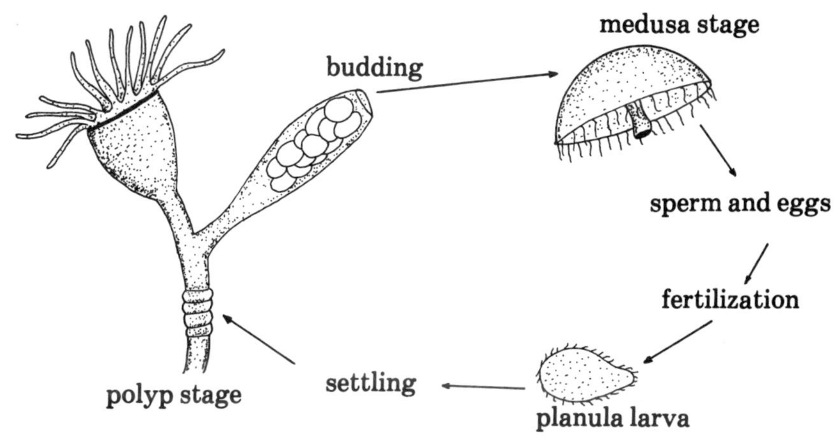

14. Generalized hydrozoan life cycle

Class Hydrozoa

the hydroids and hydromedusae

Hydroids are generally delicate colonies composed of a few to hundreds of individuals (called **zooids**). Not all of the individuals in the colony feed, some (called **gonozooids**) being especially adapted for the production of medusae, while others function as defenders using batteries of nematocysts to protect the colony from small predators. Such defensive individuals are called **dactylozooids**. The feeding zooids (called **hydranths** or **gastrozooids**) obtain food which is transported throughout the colony via a continuous gut tube. The presence of several forms of individuals is called **polymorphism**. Figure 15 illustrates a small piece of an *Obelia* colony showing both feeding and reproductive zooids. A hand lens or low-power microscope is necessary to see the details and beauty of these delicate creatures. However, with good light, and with the animals under water, you can see the tiny zooids with the naked eye when they extend their long tentacles. The overall form of hydrozoan colonies varies greatly, and immediate recognition of one is often difficult. Look for drab, bushy, branched or feathery colonies attached out of the direct force of the waves to protected rocks, under ledges, or on pilings and floats in bays. Common genera on the north coast include *Obelia*, *Eudendrium*, *Aglaophenia*, and *Tubularia* (figs. 15-

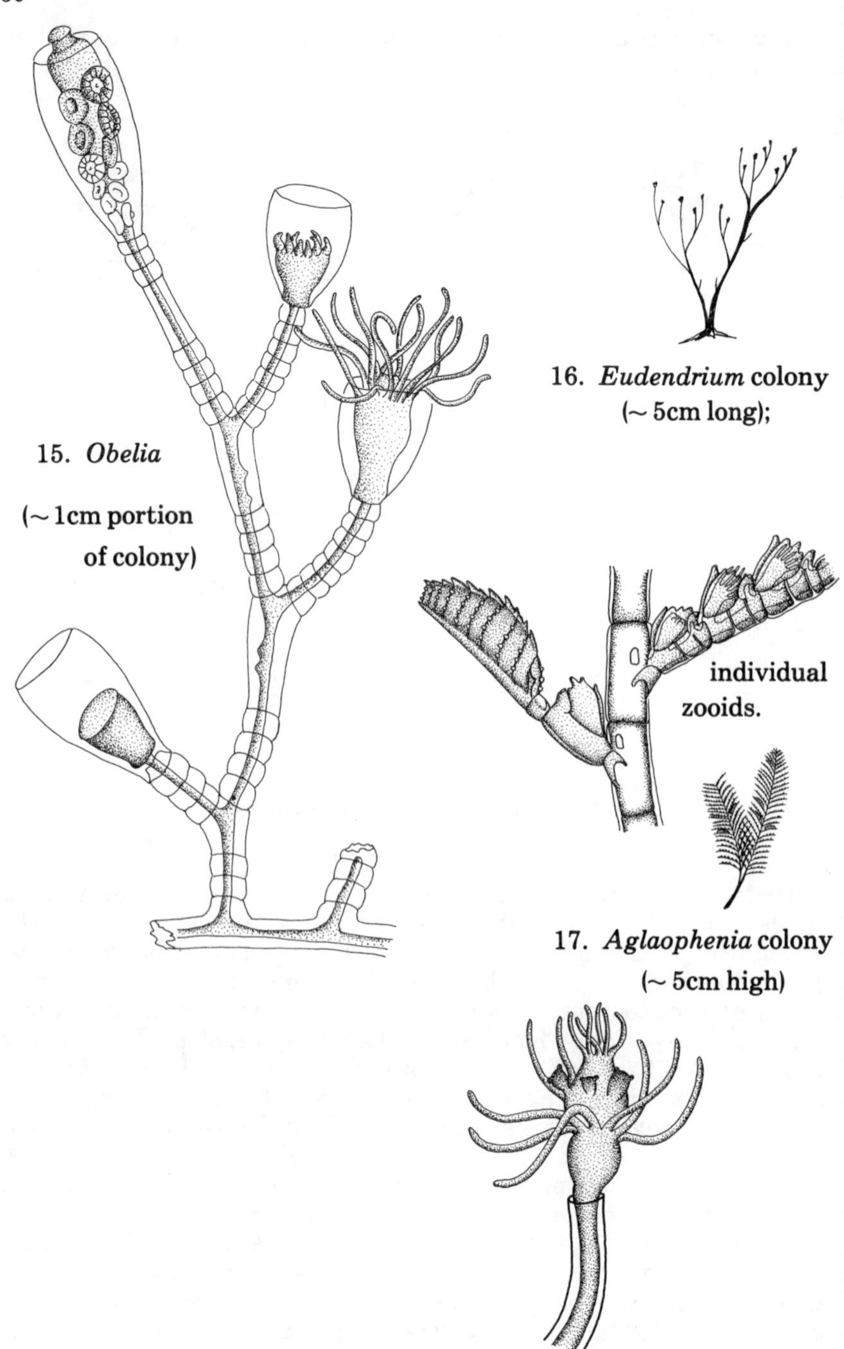

15. *Obelia*

(~ 1cm portion of colony)

16. *Eudendrium* colony (~ 5cm long);

individual zooids.

17. *Aglaophenia* colony (~ 5cm high)

18. *Tubularia*, one individual (~ 1cm high)

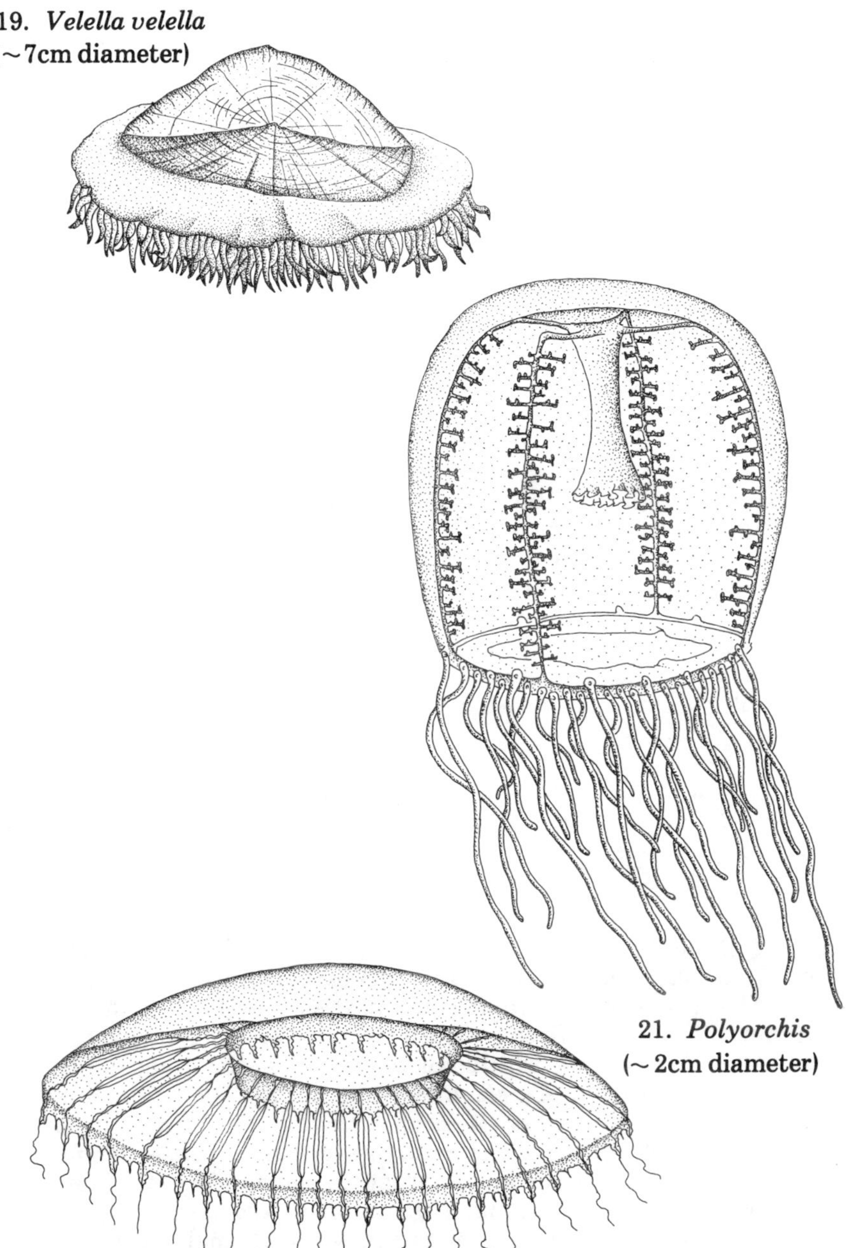

19. *Velella velella* (~7cm diameter)

21. *Polyorchis* (~2cm diameter)

20. *Aequorea* (~12cm diameter)

18). Not all hydrozoans are sessile, some are bizarre floating colonies generally found offshore. The Portugese Man-O-War (*Physalia*) is a well known tropical example of this type of hydrozoan. One such creature, *Velella velella* (fig. 19), may occasionally be found washed ashore along the north coast in large numbers. Commonly called the "by-the-wind-sailor", the bluish *Velella* is often mistaken for a stranded jellyfish, but is actually a planktonic hydrozoan colony which normally drifts floating on the water and pushed along by wind blowing against its sail, while the living individuals hang beneath the sea surface. The sail is oriented in such a way that the prevailing northwesterly winds tend to push the animals away from shore. Thus, it is following periods of unusual weather and changes in wind direction that these creatures are blown landward and become stranded.

A great variety of medusoid stages of hydrozoans (hydromedusae) may be found either washed ashore or swimming in the shallow waters of Humboldt Bay and other similar environments. A stroll through the eelgrass beds on an early-morning summer low tide will usually reveal several types to the careful observer. Many of these medusae may also be seen from floating docks in Humboldt and Trinidad Bays. Except for the large *Aequorea* (fig. 20) and *Polyorchis* (fig. 21), most local hydromedusae are small, less than 2-3cm across, and a hand lens is necessary to see much detail and make possible specific identification.

Class Scyphozoa

the scyphomedusae or "true" jellyfishes

All of the very large jellyfishes are pelagic, drifting in the near and offshore waters, and only seen by non-seagoing observers when they have been washed ashore or come into shallow bays with the tide. Again, the bay sand flats are good places to search for such stranded animals. Two species which are frequently encountered locally are the clear dish-shaped *Aurelia aurita* (fig. 22) and the large brown and sometimes purple *Chrysaora melanaster* (fig. 23). The latter species bears relatively powerful nematocysts and may cause itching or a slight rash when handled, even when the animal is dead. Some particularly sensitive people may suffer more severe allergenic reactions. A close inspection of these animals may reveal small red shrimp-like crustaceans clinging to the cnidarians' surface. These part-time hitch-hikers are hyperiid amphipods of the genera *Hyperia* and *Hyperoche*, animals normally found in the open ocean, which apprarently feed on the tissues of their hosts.

One local scyphozoan, *Haliclystus* (fig. 24), does not resemble a typical jellyfish at all. This strange animal is intertidal and sessile

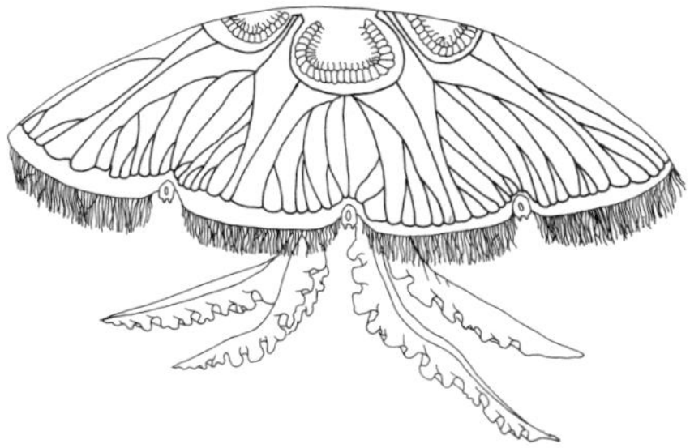

22. *Aurelia aurita*, top view (~ 10cm diameter)

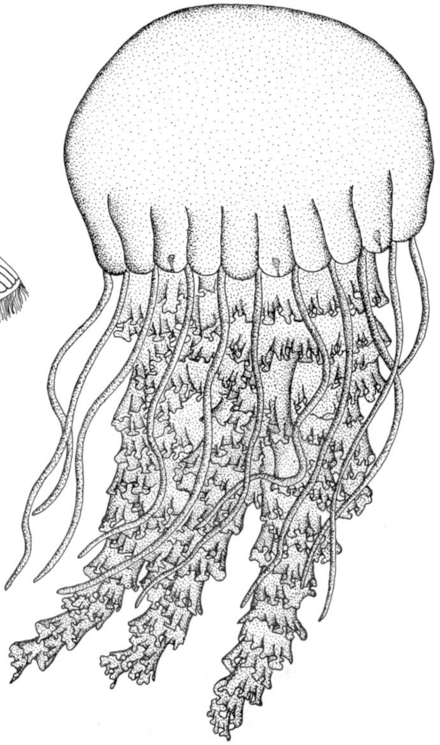

23. *Chrysaora melanaster* (~ 20cm diameter)

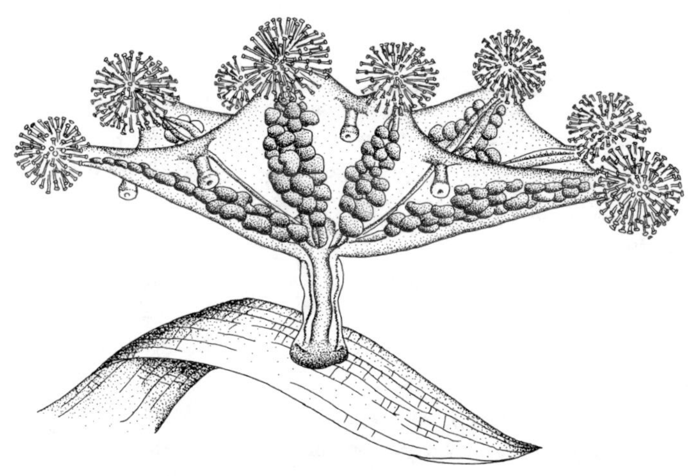

24. *Haliclystus* (~ 1cm high)

and may be found attached upside down to seaweeds in quiet tidepools or to eelgrass in Humboldt Bay. Although *Haliclystus* looks more like a polyp, this small (about 1cm high) flower-like animal is actually an attached medusa. True scyphozoan polyps are very tiny inconspicuous stages in their life cycles and difficult to see without the aid of a hand lens. The polyps of *Aurelia* are abundant in Humboldt Bay, particularly on docks and piers, but are usually overlooked by the casual observer. The feeding phase of the polyp (called a **scyphistoma**) resembles a miniature white anemone, about 2mm long, and the budding stage (called a **strobila**) is simply a more elongate form (up to 2cm long) whose body buds off tiny medusae.

Class Anthozoa

the sea anemones and corals

Some of the sea anemones are among the largest and most conspicuous intertidal animals along our coast. Although much larger than the zooids of a hydroid colony, anemones are constructed in the same general way. Those species included in the key and discussed below can all be found with little effort.

Key to Some Common North Coast Anemones

1. Body deep red or orange, sometimes mottled with green or white 2
1. Body green, pink, brown, grey, or white, or some combination of these colors 4
2. Small (generally less than 2cm across) bright orange bodies; animal encased in a hard, calcareous outer covering (feel the animal!) .... *Balanophyllia elegans* (fig. 30)
2. Large (often several centimeters across); body deep or bright red, usually mottled with green or white; body not encased in hard skeleton 3
3. Body with dull green markings ...... *Tealia crassicornis* (fig. 29)
3. Body with white markings .... *Tealia lofotensis*
4. Body smooth; white, grey, or brown; with very numerous short thin tentacles; usually on pilings or floats in quiet water .... *Metridium senile* (fig. 25)
4. Not as described or figured above 5

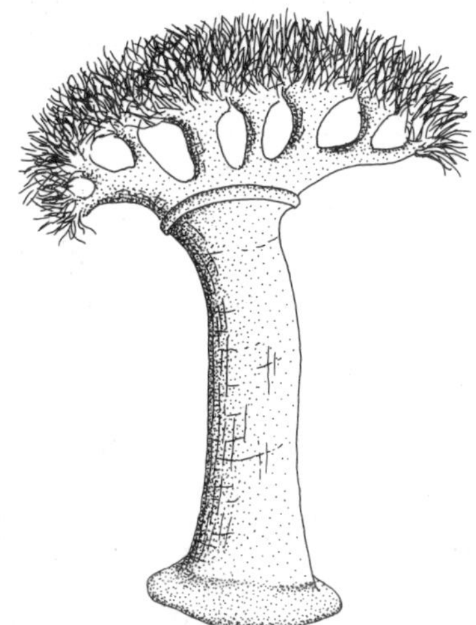

25. *Metridium senile*, fully expanded (~ 7cm high)

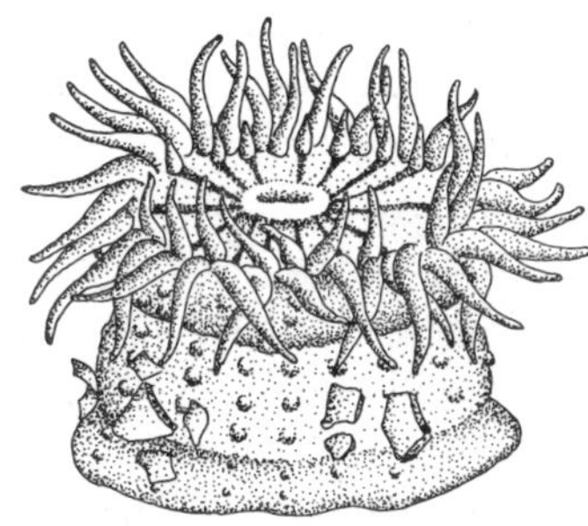

26. *Anthopleura elegantissima* (~ 3cm diameter)

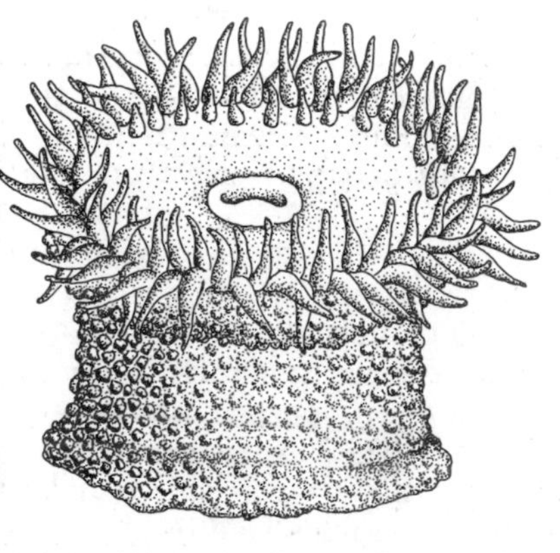

27. *Anthopleura xanthogrammica* (~ 12cm diameter)

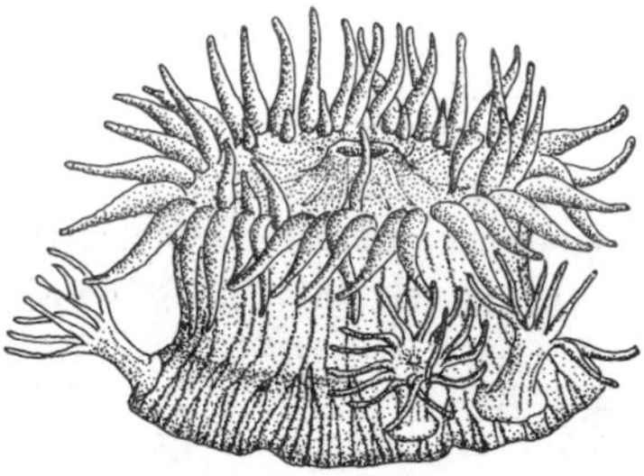

28. *Epiactus prolifera* with young attached to the base (~ 2cm diameter)

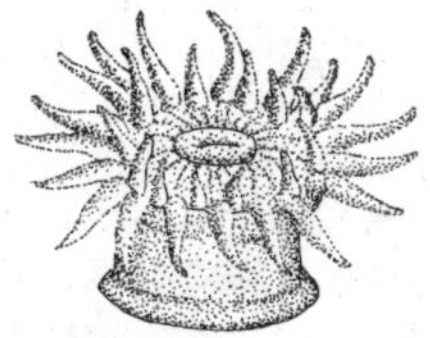

30. *Balanophyllia elegans* (~ 1cm diameter)

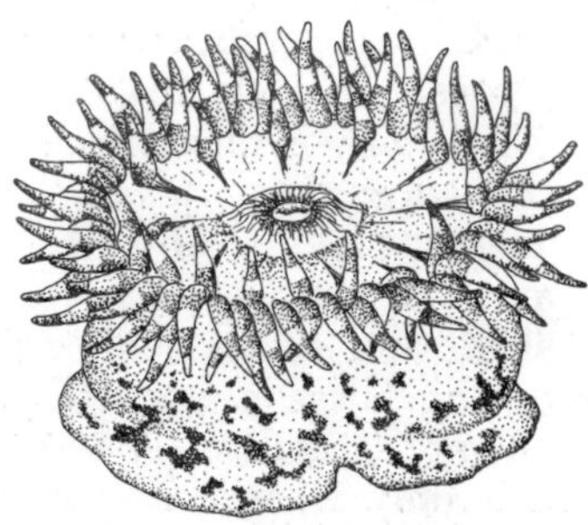

29. *Tealia crassicornis* (~ 7cm diameter)

5. Body smooth; usually greenish with lighter stripes across the upper end radiating outward from the area of the mouth; often with young anemones attached to the base; usually on brown seaweeds (less frequently on the undersides of rocks) .... *Epiactis prolifera* (fig. 28)
5. Not as described or figured above 6
6. Small (usually 2-5cm across); tentacles pink, body grey-green; usually found in dense aggregations on upper-mid-intertidal rocks .... *Anthopleura elegantissima* (fig. 26)
6. Large (often 10cm or more across) solitary anemones; usually bright green; found in lower rock pools or on pilings................................... *Anthopleura xanthogrammica* (fig. 27)

The most commonly seen sea anemone in Humboldt and other northern California bays is *Metridium senile* (fig. 25). Members of this species range in color from white to nearly chocolate brown. Aggregations are generally dominated by a single color phase, a result of all the individuals having come from a single parent by asexual reproduction (a process known as **cloning**). Most of these animals occur in the low intertidal zone, or on floats and pilings which are not exposed during most low tides. *Metridium* usually does not exceed 10cm in diameter, and most are much smaller. This species is easily identifiable, when it is expanded, by the very high number of thin tentacles, giving the animal a frilled or fuzzy appearance at the mouth end. Having a high number of small tentacles allows this animal to capture small beasts and particles from the water as well as taking large prey.

Two species of *Anthopleura* are abundant along the California coast. The rocks in the upper-mid-intertidal zone of semiprotected shores are the favorite haunts of the "aggregating" anemone, *Anthopleura elegantissima* (fig. 26). In some areas the rock surfaces may be completely covered by these animals, packed so closely together as to form a solid mat of anemones. They are generally contracted when exposed by low tides, and their bodies are usually covered by bits of shell fragments and small pebbles held in place by sticky secretions. These small anthozoans (up to a few centimeters across) are greyish to greenish in color, with bright pink tentacles which are evident when the animals expand under water. *Anthopleura elegantissima* lives in a relatively high tidal zone for sea anemone and is subjected to frequent and often prolonged exposure to air. The habit of forming dense aggregations plus the protection of shells and stones held to their surfaces allow the animals to stay wet during these periods. *Gently* press your hand against the mat of anemones and water will flow as though from a soaked sponge. Do this only when the incoming tide is near to the animals, otherwise they may dry out and not survive. These habits also help protect the animals from physical damage from waves or floating objects to their fleshy bodies. Do not walk on them, however, as they are not that tough! Aggregating probably also facilitates the capture of larger prey than would be possible by a solitary anemone.

The large beds of *A. elegantissima* are actually clones, consisting of individuals derived from a single progenitor by asexual means (as described above for *Metridium*). Individual clones do not mix but remain separated from adjacent clones by a thin strip of bare rock acting as a sort of "no-man's-land". In fact, neighboring clones actually engage in aggresive behavior with special nematocyst-bearing structures near the bases of the tentacles to maintain the separation.

Anthopleura xanthogrammica (fig. 27), the giant green anemone, lives from the mid- to low-intertidal zones. It is more solitary in its habits than *A. elegantissima* and not nearly as resistant to aerial exposure. As its common name indicates, the animal is large (sometimes 20cm in diameter) and generally bright to pale green in color. Some of the green color is due to symbiotic algae which live deep in the tissues of the anemone. You may notice that those anemones which live in areas which receive little or no sunlight are paler than those exposed to sun more frequently. Rocky tidepools, surge channels, and deep pilings are some common habitats of this species of *Anthopleura.*

A third species, *Anthopleura artememisia* (not in key), is a partial burrower, and resembles *A. elegantissima* in the possession of pinkish tentacles. Look for *A. artemisia* in sandy areas where stones are present beneath the sediment. The animal attaches to these stones and extends its tentacles up into the overlying water.

Epiactis prolifera (fig. 28) is one of the most attractive and interesting species of anemones on our coast. Generally from 2-4cm in diameter, *Epiactis* is usually translucent deep green (occasionally brown) in color, often with light stripes across the upper surface and sometimes down the sides of the body column or trunk. The tentacles may have lighter, nearly whitish patches of color scattered along their length. These animals frequently bear numerous small anemones appearing like miniature adults attached to the base of the parent where they have developed from embryos. These young anemones eventually crawl away to establish a solitary adulthood. *Epiactis* is often seen on the blades of brown seaweeds in the low tide zones, or on the undersides of rocks.

Two species of large, distinctly red anemones may be found in the lower tidal zones. *Tealia crassicornis* (fig. 29) is rich red in color, with mossy-green markings, while *T. lofotensis* is bright red with white markings. Look for these animals in the deep low crevices along jetties in Humboldt Bay, or in well-protected rocky areas.

One species of true hard coral is common along some of the low intertidal rocks from Monterey to Puget Sound. Locally it is particularly abundant at Trinidad. *Balanophyllia elegans* (fig. 30) only occurs in shadowed areas under ledges, out of the direct force of waves and sunlight. These animals are not reef-building corals and are found either singly or in small aggregations. By touching them you can cause the soft body to retract, exposing the hard "coral" exoskeleton.

PHYLUM CTENOPHORA (comb bearers)

the comb jellies and their kin

Ctenophores are planktonic, open-ocean animals, not normally encountered in the tidal zones. One common northeastern Pacific species (*Pleurobrachia bachei*, fig. 31) may be found washed ashore along sandy beaches or drifting in Humboldt and Trinidad Bays. Frequently the washed-up ones are alive and, when put in a jar of sea water, swim and extend their delicate tentacles. These animals appear as small spheres of clear jelly (about 1cm diameter) and are commonly called "cat's eyes" or "sea gooseberries". They are actually complicated creatures which are voracious predators in the planktonic food web and will consume nearly any animal of proper size which they can detect and hold. A pair of long retractible tentacles may be seen in intact specimens, each bearing mucus-producing cells (**colloblasts**) for prey capture. These tentacles may actually be "fished" by various maneuvers such as swirling the body, somersaulting, etc. The animal propels itself by the action of eight rows of tiny flickering paddles called **ctenae** (combs) from which the name of the phylum is derived. These creatures are generally considered to be related to the cnidarians, probably the Hydrozoa, and represent another major group of diploblastic animals. Ctenophores display biradial symmetry as described in an earlier chapter. Incidentally, *Pleurobrachia* frequently harbors little red amphipods (*Hyperoche*) in an association similar to that noted previously for certain scyphomedusae.

PHYLUM PLATYHELMINTHES (flatworms)

Class Turbellaria

the free-living flatworms

The Phylum Platyhelminthes includes the tapeworms (Class Cestoda), the flukes (Class Trematoda), and the "free-living" (i.e. not parasitic) flatworms (Class Turbellaria). While there are many examples of the two groups of parasitic types, and you may find them by picking through the innards of fishes, our concern along the seashore is with the turbellarians. The flatworms possess more complicated bodies than any of the phyla discussed thus far. The most obvious difference is that the flatworms are bilaterally symmetrical and move about freely, head first, in search of food and whatever else flatworworms seek. Most of their sense organs are concentrated in the head end, although the mouth is usually set well back along the ventral

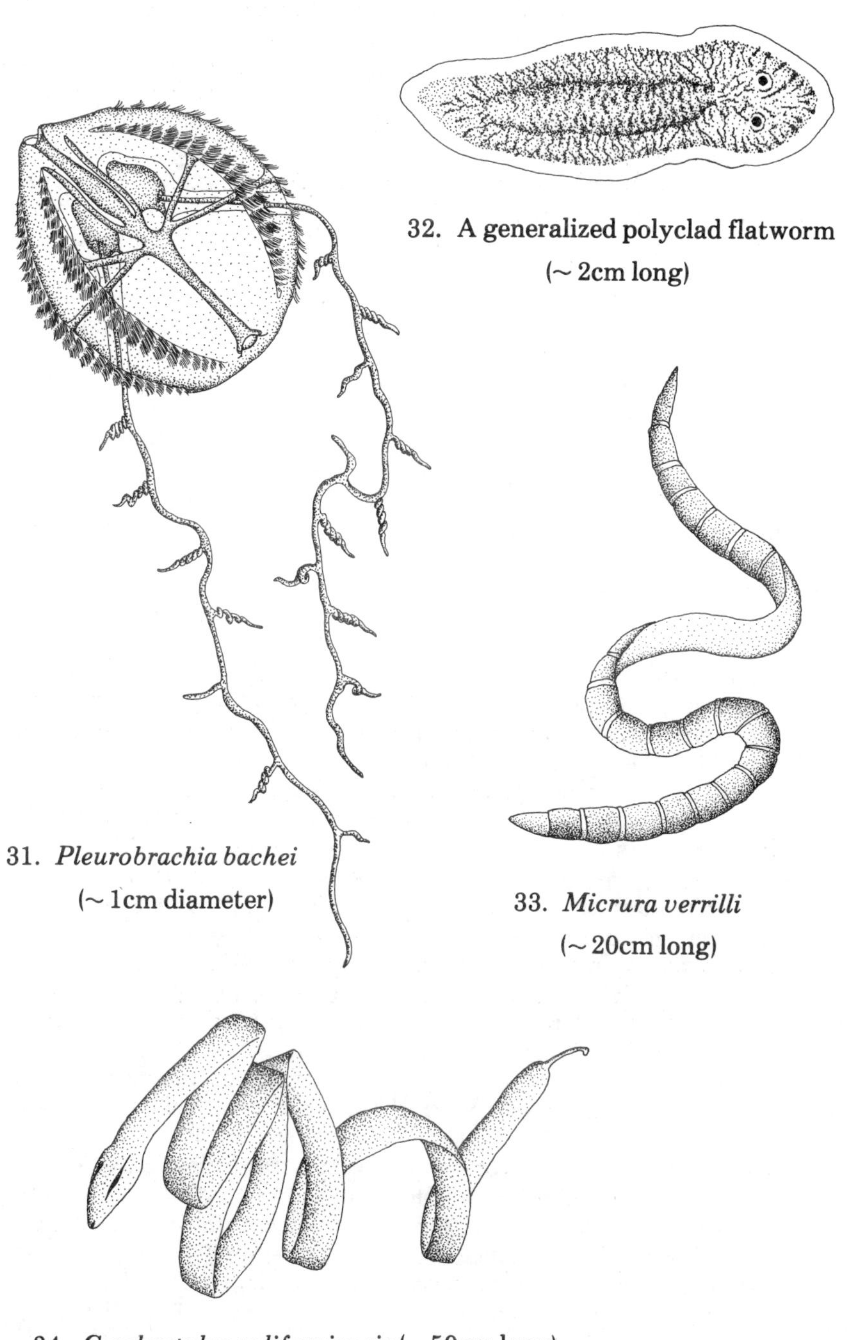

32. A generalized polyclad flatworm (~ 2cm long)

31. *Pleurobrachia bachei* (~ 1cm diameter)

33. *Micrura verrilli* (~ 20cm long)

34. *Cerebratulus californiensis* (~ 50cm long)

surface. In addition to this difference in symmetry, flatworms are triploblastic, and the addition of true mesoderm provided an additional tissue layer allowing, in part, the development of the complex musculature necessary for their coordinated unidirectional movement, and the advent of very complicated reproductive systems. These acoelomate creatures also have incomplete digestive systems, with a single opening serving as both the mouth and the "anus". Turbellarians are **hermaphroditic**, that is, each individual contains both male and female reproductive systems. When mating they mutually exchange sperm, then each mate goes about its business acting as an impregnated female. Many species have a free-swimming larval stage which, if it survives, settles and becomes another worm. It is worthwhile to mention here that the production of free-swimming larvae by various invertebrates offers certain distinct advantages. These advantages are especially important to those animals which are benthic and sessile or sedentary during their adult lives. The larvae are carried and dispersed by currents and tides and can subsequently settle and become established in new available areas. Many larvae feed and, because of their planktonic habits, utilize a different food source from that of the adults, thus eliminating competition between the two phases of the life history. You can probably think of other advantages (and perhaps some disadvantages) to the presence of planktonic larvae in an animal's life cycle.

Most intertidal flatworms are drab brown or grey, and a generalized one is shown in figure 32. One of the most common genera in these parts is *Notoplana*, which resembles closely the figured individual. These are inconspicuous worms up to about three centimeters in length, and are commonly found in cobble areas, on the undersides of rocks in the mid- to low-intertidal zone, or crawling over mussel beds on foggy mornings. They are one of the few beasts adapted to live in rather unstable rock areas. These oval-bodied worms are so flat and deformable that they can live in tiny depressions in rocks and thus are not crushed if the rock is rolled about by waves. Flatworms glide along the substrate by means of microscopic hair-like cilia that are concentrated on their undersides, aided by udulations of the sides of the body. Most turbellarians are predators on tiny soft-bodies animals or various sorts. To identify specifically members of the Turbellaria, one must make a detailed microscopic examination of the sexual parts, a job we generally leave to those with the appropriate time and inclination for such things.

Not all "free-living" flatworms are, in fact, free-living. *Syndisyrinx franciscanus* is a turbellarian which inhabits the guts of local sea urchins (*Strongylocentrotus*). This small reddish worm has rather transparent skin, providing an excellent view of the complex reproductive systems (if one has a proper microscope).

PHYLUM NEMERTEA (unerring)

the ribbon worms

Ribbon worms are considered to be close relatives of the flatworms because of similarities in construction and embryogeny between the two groups. On the other hand, the nemerteans show some major increases in complexity over their presumed ancestors. For example, the ribbon worms posses a complete gut, with mouth and anus at opposite ends of the body, and they have a complicated blood-circulatory system that is entirely lacking in flatworms. Most nemerteans are **dioecious** (having separate sexes) and their reproductive systems are surprisingly simple, consisting of transient mesodermal sacs in which the sperm and eggs form. A planktonic larva provides a short free-swimming period in the worm's life cycle, prior to settling.

Many ribbon worms are brightly or distinctly colored, often with patterns which are diagnostic of the particular species. Their long thin bodies are extremely rubbery and extensible, capable of being stretched to several times their contracted length (some may be 2-3m long when fully extended). These beautiful worms usually possess an eversible snout, or **proboscis**, typically separate from the mouth but emerging close to it, which may be armed with piercing spines and used for the capture of prey. They feed upon soft-bodied invertebrates, mostly polychaete worms. The species included in the key and discussed below are relatively common along the northern California coast and/or have distinctive color patterns which aid in identification.

Key to Some Common North Coast Ribbon Worms

1. Body flat, relatively short and broad (up to about 8cm long); found within the mantle cavities of bay clams ...... *Malacobdella grossa*
1. Body very elongate, thin; free-living 2
2. Body bright orange, very long (up to 2-3m) .... *Tubulanus polymorphus*
2. Body not bright orange 3
3. Body white to pale pink, without any distinct color pattern; small thin worms (up to about 15cm long); common in mussel beds and algal mats .... *Amphiporus* sp.
3. Body not overall white or pink 4
4. Body rather thick and fleshy, long (up to 1m); local specimens are uniformly dark brown; mostly found in tidal flats .... *Cerebratulus californiensis* (fig. 34)
4. Not as described or figured above 5
5. Body dull green, small and thin (up to about 15cm); common in

mussel beds and on protected pilings and floats .... *Emplectonema gracile*

5. Body not green 6
6. Body solid purple to purplish-brown on back, up to about 25cm long; with a distinct white marking at back of head .... *Paranemertes peregrina*
6. Body with deep purple blocks of color separated by white bands across the back; up to about 40cm long .... *Micrura verrilli* (fig. 33)

If you are a clam-digging enthusiast in Humboldt Bay it is likely that you have already encountered one ribbon worm. *Malacobdella grossa* is a symbiont living withing the mantle cavity of some local clams. It has become adapted for this life style by a reduction in sense organs and the acquisition of a more compact shape than that of most other nemerteans. *Malacobdella* is considered a **commensal** since it benefits from its association with the clam, yet causes no obvious harm to its host. This worm feeds on the detritus and microscopic organisms in the water currents passing through the clam's gill cavity.

The free-living nemerteans are all long and thin. Among the most visible and beautiful of these is the bright orange *Tubulanus polymorphus*. Members of this species reside throughout northern California and Oregon in the mid to low intertidal zones on rocky coasts. They may reach lengths of three meters or more. Overcast mornings when the tide is very low are the times to look for this and other nemerteans prowling about the algal-covered rocks and low pools. The rocky areas south of Patrick's Point include some fine habitats for *Tubulanus*. The northern author has stood in a large pool and observed no less than a half dozen of these lovely worms at one time.

Several other ribbon worms may be encountered in the rocky intertidal area. *Micrura verrilli* (fig. 33) is most commonly seen on the undersides of rocks in semiprotected areas from Monterey to Alaska. Look particularly on stones which are resting in sand for this often overlooked worm. *Micrura* is easily recognized by the distinctive purple and white pattern along the back (**dorsum**). It may attain a length of about 40 cm, but most are somewhat smaller.

Amphiporus, *Paranemertes peregrina*, and *Emplectonema gracile* are separble on the bases of their coloration (see the key). They may all be found on foggy mornings crawling over the surface of mussel beds and protected pilings (*Paranemertes* may also be seen on the surface of mud flats). An ideal place to view scores of *Emplectonema* is on the underside of the wooden boat ramp near Samoa on the north spit of Humboldt Bay. As this book is being written, however, the old ramp is being destroyed and replaced with one of concrete. We are not certain how much, if any, of the old structure is going to be left intact, but we will be saddened indeed if it is to be fully dismantled and joins the scrap heaps. If it is not too late as you read this, you are encouraged to visit beneath the ramp at low tide and see an incredible array of creatures not normally encountered intertidally.

Digging in the tidal flats of Humboldt Bay may turn up one of the largest ribbon worms along this coast, *Cerebratulus californiensis* (fig. 34). The rather thick brown body appears quite robust, but the animal readily fragments when handled too roughly. The flattened, spade-shaped head possesses a deep slit along each side with which the animal senses chemical changes in its environment and may detect prey.

PHYLUM ANNELIDA (annulated, ringed)

the segmented worms

The Phylum Annelida includes the Classes Polychaeta, Oligochaeta, and Hirudinea. The latter two are composed of the earthworms and their relatives, and the leeches respectively. While there are marine examples of these two groups, the polychaetes are by far the most numerous and important annelids in the sea. They include a bewildering array of body forms and habits, from **errant** (actively motile) predators and herbivores with heavy jaws, to sedentary tube dwellers which filter food from the water with feathery tentacles. There are well over 5,000 described species of polychaetes, and new ones are being discovered and described at an overwhelming rate. Of the many species which inhabit the north coast only a few examples are treated here. As their common name indicates, annelids are worms whose body is divided into rings or segments, a feature which quickly distinguishes them from all other marine worms. Each segment bears a pair of appendages called **parapodia** which may assume a great variety of shapes and functions. The parapodia are equipped with tiny spine-like projections called **setae**, the shapes of which are of considerable importance in specific identification. Many polychaetes are commonly called "bristle worms" in reference to the numerous setae along the sides of the body.

Internally the annelids are complex. Their complete gut is usually specialized for the particular kind of food consumed and thus varies more in structure from species to species than does that of flatworms or ribbon worms. The segmented worms have a well-developed circulatory system and a more complicated set of excretory structures than either the platyhelminths or nemerteans. In addition, annelids are coelomate; each segment of the main portion of the body (the trunk) contains a pair of coelomic spaces, one on each side of the digestive tube, each with walls (called **septa**) partitioning off one segment from the next. These fluid-filled sacs allow patterns of body movement and support for the internal organs not possible for acoelomate animals. In the polychaetes, and many other coelomate animals, the coelomic spaces house the gametes which arise from the mesodermal linings. All in all, the body construction of annelids is far more complicated than that of the previously discussed groups, and this complexity is partially responsible for the great diversity and success of the members of the phylum.

Most polychaetes are dioecious, and the gametes are usually released into the sea water where fertilization occurs. Many have a patter of development which results in the formaton of a very characteristic, top-shaped, swimming larva called a **trochophore**. The larva progressively adds segments as it elongates, eventually settling as a little worm. Trochophores come in a variety of shapes and sizes, and are often identifiable to at least the level of family.

Polychaete identification is often extremely difficult except for a handful of very distinctive species, and few workers can claim real expertise in this matter. Do not be disturbed when you find species that are not included in this handbook, you are reminded that only a small fraction of the total number of local polychaetes is mentioned here. Those chosen for inclusion are generally large, conspicuous, common, or easily identified in one way or another. In addition we have tried to use examples which illustrate the great adaptive range within the class.

Key to Some Common Types of Polychaetes

1. Body thick and oval; dorsum covered with long, dense, hair-like setae .... *Aphrodite* sp. (fig. 35)
1. Body form variable, but not as above, usually somewhat elongate; dorsum not covered with dense hair-like setae 2
2. Dorsum of body more or less covered with flattened overlapping scales (figs. 36-37); the scale worms (Family Polynoidae) 3
2. Without obvious scales on the dorsum 5
3. Worms living in grooves on undersides of seastar arms, or beneath the shell edges of large limpets, or in the gill grooves of chitons .... *Arctonoe* spp. (fig. 36)
3. Not living with seastars or mollusks as described above 4
4. Found in bay mud, in the burrows of the "fat innkeeper" *(Urechis)* ....*Hesperonoe adventor*
4. Free-living, relatively common.............................. *Halosydna brevisetosa* (fig. 37) (and other free-living scale worms)
5. In hard, white, irregularly twisted, calcareous tubes; usually attached to rocks, pilings, or shells; tentacles generally bright red .... Family Serpulidae (fig. 42)
5. Free-living, or burrowing, or living in sand or parchment-like tubes 6
6. Living in tough parchment-like tubes attached to solid substrates, or in rigid sand tubes in soft substrates 7
6. Free-living, burrowing, or in tubes other than described above ... 9

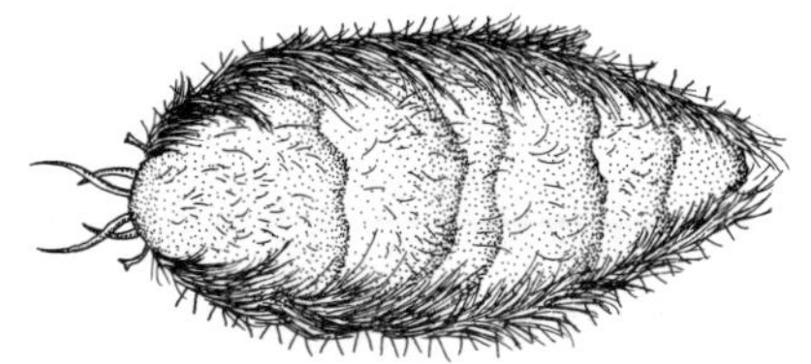

35. *Aphrodite* (~5cm long)

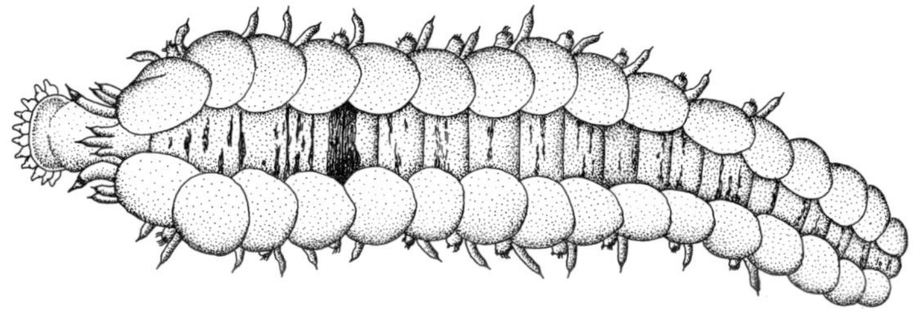

36. *Arctonoe vittata* (~3cm long)

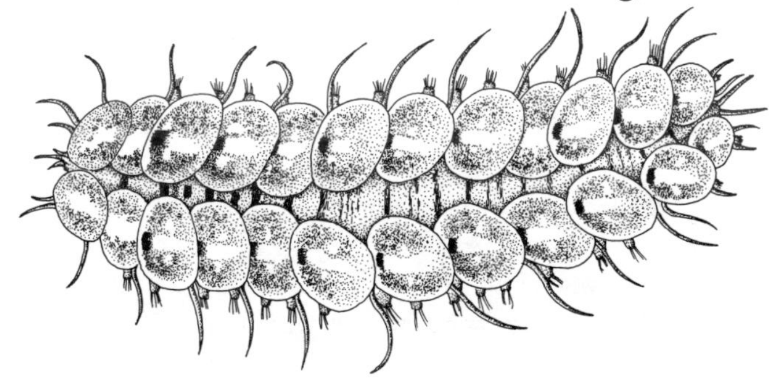

37. *Halosydna brevisetosa* (~5cm long)

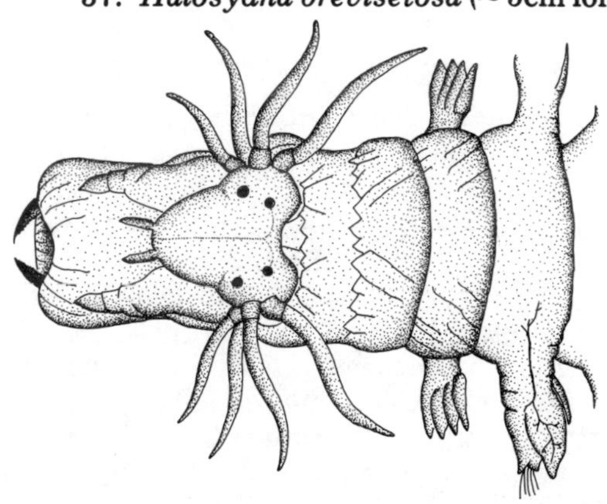

38. *Nereis* head with everted proboscis showing jaws
entire worm ~15cm long

7. Living in elongate, sandy cone-shaped tubes. . . . . *Pectinaria californiensis*
7. Living in parchment-like tubes 8
8. Occuring singly in sand flats of bays; tube with fringed hooded end protruding a few centimeters above sand surface. . . . . *Pista pacifica* (fig. 40)
8. On docks and floats; in clusters of non-hooded tubes; with feathery red, green, or purple tentacles . . . . *Eudistylia* spp. (fig. 41)
9. Parapodia flap-like, relatively large, and generally similar to one another along the length of the body; usually active worms . . . . Families Nephtyidae, Glyceridea, Nereidae, Syllidae, Phyllodocidae, and others, see figures 38-39; further identification is beyond the scope and purpose of this handbook.
9. Parapodia either small and not flap-like, or if flap-like distinctly separable into more than a single type along the length of the body; usually sedentary and burrowing 10
10. Living in a U-shaped tube buried in tidal flats; body formed into distinct regions and with large fan or flap-like parapodia . . . . *Mesochaetopterus taylori*
10. Not in a U-shaped burrow; parapodia not large and flap-like . . . 11
11. With numerous long thread-like "tentacles" along the sides of the body . . *Cirriformia spirabrancha* (and other members of the Family Terebellidae)
11. Body generally without obvious tentacles or appendages, greenish; burrowed in sandy, semi-protected areas . . . . . . *Abarenicola* sp.

Members of the genus *Aphrodite* (fig. 35) are most frequently found subtidally, but on occasion they may be seen on soft sediments within the low-intertidal zone, or washed ashore after a storm. The common name of "sea mouse" is based on the rather hairy appearance of these worms when viewed from the top; the underside, however, reveals the distinctly segmented nature of the body. The "hairs" are in fact modified parapodial setae. In many species these dorsal setae have a beautiful golden or coppery sheen. These worms should be handled with caution as the setae are sharp and can inflict painful wounds.

The scale worms (Family Polynoidae) have their dorsal surfaces more or less covered by scale-like portions of the highly modified parapodia (figs. 36-37). Some of these worms (usually the smooth-scaled species) are found in close association with other invertebrates, while others are typically free-living (usually those with ornamented scales).

Two relatively common species of commensal scale worms occur along the northern California coast. They normally inhabit the grooves on the undersides of the arms of seastars, or the space beneath the shells of gastropods between the edge of the shell and the foot (especially the keyhole limpet, *Diodora*, fig. 80), or in the gill chambers of the giant gumboot chiton, *Cryptochiton* (fig. 72). Both of these worms belong to the genus *Arctonoe*, and both are basically white in color. *Arctonoe vittata* (fig. 36) has a dark-brown to nearly black band across its back near the head end which distinguishes it from its less common relative *A. fragilis*. These animals are not parasitic, but rather feed upon bits of detritus which enter their protected homes.

The burrows of larger invertebrates often provide lodging for a number of smaller animals. *Hesperonoe adventor* is a reddish-brown scale worm which inhabits the burrows of the "fat innkeeper," *Urechis caupo* (fig. 44). *Hesperonoe* picks at bits and pieces of food drawn into the burrow by the water current created by its host. There are other boarders in the burrow of *Urechis* including a crab and a small fish (see *Urechis*). Perhaps the most common free-living scale worm along our coast is *Halosydna brevisetosa* (fig. 37). This species (and other scale worms) may be found within the tubes and burrows of other polychaetes but is more commonly seen in the protected understories of mussel beds and the undersides of rocks.

Errant polychaetes which actively hunt prey or graze on algae are common and represent several different families (for example Nereidae and Glyceridae, figs. 38-39). These groups include some large worms which may be seen in a variety of habitats. Most of the errant polychaetes have all of the body segments more or less similar to one another (and are thus said to be **homonomous**), well-developed parapodia which aid in locomotion either by swimming or crawling over the substrate, and an internal jawed **proboscis** which can be everted for capturing prey or tearing off pieces of algae. *Nereis* (Fig. 38) is common under stones and in mussel beds; *Glycera* (fig. 39) is more frequently seen in soft substrates into which it burrows. Members of both genera feed largely on other soft-bodies animals, although we have seen some of these so-called predators bite off pieces of algae as we might eat an apple. There is still much to learn in regards to feeding habits of worms. The proboscis of *Glycera* is particularly long and bears four sharp hook-shaped jaws. You may be able to induce the eversion of the proboscis by gently stroking the animal along its back. Most of the errant predators attain lengths to 15 to 20 centimeters. The largest, however, is a nereid of the genus *Neanthes* which occurs in Humboldt Bay and may reach more than two meters in length!

Many polychaetes have given up the wandering life-styles described above to become sedentary tube dwellers or relatively inactive burrowers. The bodies of these worms are usually regionally specialized into various sections (**heteronomous**). This condition allows the worm to carry on a number of activities at the same time, so to speak. One body region may possess bristles to hold the animal in place within its tube, while another section may undulate and pump water over

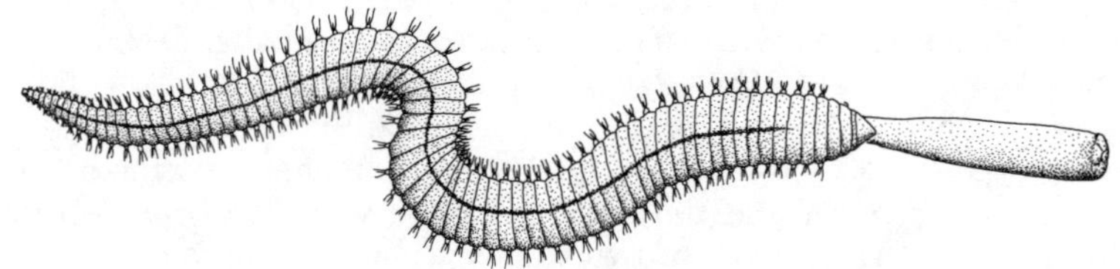

39. *Glycera*, with proboscis extended (~ 15cm long)

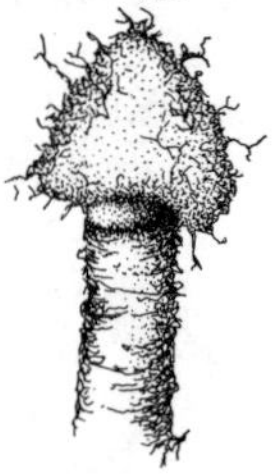

40. The characteristic tube of *Pista pacifica*

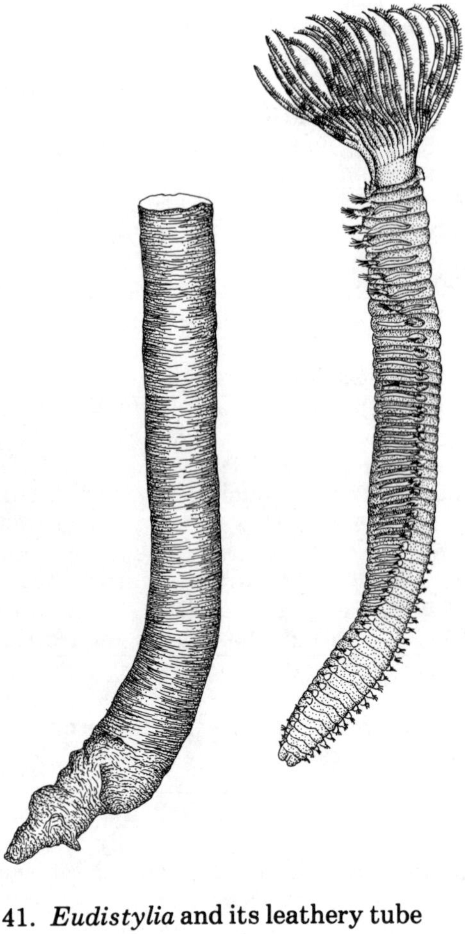

41. *Eudistylia* and its leathery tube (~ 15cm long)

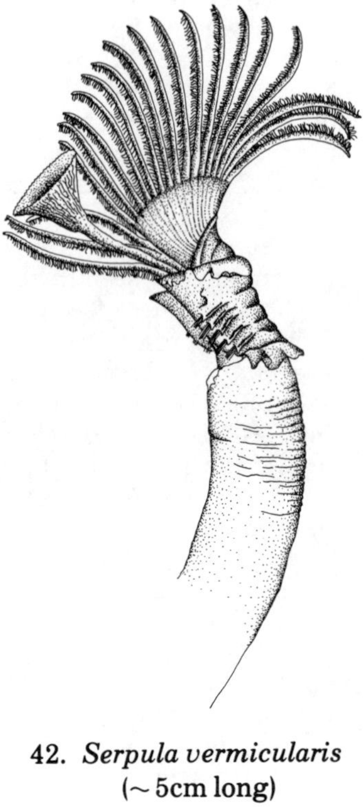

42. *Serpula vermicularis* (~ 5cm long)

the feeding area and gills. Feeding in sedentary polychaetes may involve filtering the sea water (e.g. *Eudistylia*, *Serpula*, *Mesochaetopterus*), ingesting sand or mud and digesting the organic material within it (e.g. *Abarenicola*), or extending sticky tentacles out into the substrate to capture detritus (e.g. *Pista* and *Cirriformia*). In any case, the sedentary polychaetes depend primarily on environmental factors to replenish the food supply in their immediate neighborhood rather than wandering about actively searching for prey.

A great variety of burrowing and tube-dwelling polychaetes can be found in Humboldt Bay. A few minutes of effort with a shovel along the low intertidal mud flats will usually turn up several such worms. Look for small openings in the mud surface or volcano-shaped mounds as evidence that these worms have burrowed into the substrate. Some of the tube-dwellers may construct their tubes vertically in the sand flats, such as *Pista pacifica* which builds a characteristic "hooded" tube (fig. 40). Others produce tubes attached to hard substrates such as docks, floats, and low intertidal rocks (e.g. *Eudistylia*, *Serpula*, figs. 41, 42).

Polychaetes are among the most numerous animals in the sea and are extremely important in the marine food web. Many are preyed upon by a great variety of vertebrates (fishes and wading birds) and invertebrates including some other polychaetes.

PHYLUM SIPUNCULA **(little siphon)**

the sipunculans or peanut worms

Sipunculans are unsegmented coelomate worms which most workers think are somehow related to the Annelida because of similarities in early embryogeny and the organization of the nervous system. The digestive system consists of a very long, coiled, U-shaped gut tube which leads from the mouth down into the main region of the body clear to the posterior end, then loops forward again so that the anus is located anteriorly on the dorsum near the base of the introvert. This anterior placement of the anus allows the animal to partially burrow and still have the anus free of the sediment. The feeding tentacles bear cilia and mucus which trap and carry tiny bits of food to the mouth. Male and female sipunculans shed their gametes into the seawater where fertilization takes place. Their early embryogeny is similar to that of polychaetes, leading to the formation of a trochophore larva which eventually elongates, without segmenting, and settles to the bottom as a tiny worm.

The common local species (*Phascolosoma agassizii*, fig. 43) has a tough leathery skin covered with small bumps and warts and marked with irregular transverse bands and spots of dark brown. The mouth is located at the end of a retractible extension called the **introvert**. Unless fully relaxed and undisturbed the introvert is retracted into the body

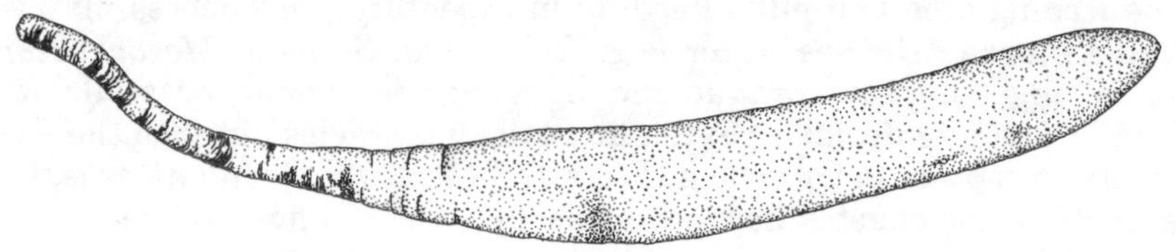

43. *Phascolosoma agassizii* (~ 5cm long)

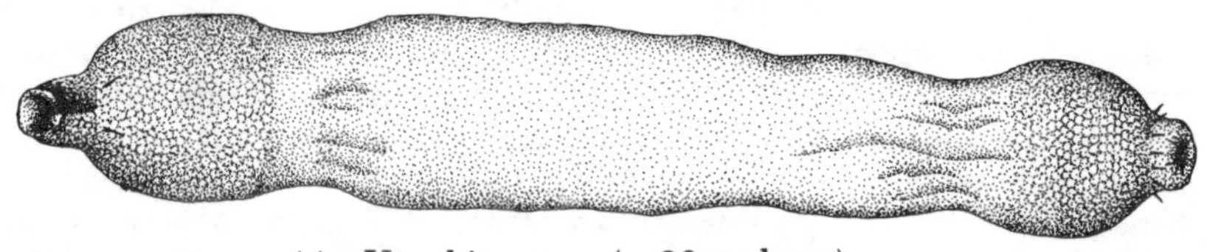

44. *Urechis caupo* (~ 20cm long)

which then has the general shape of a peanut (although many sipunculans bear no resemblance to peanuts at all). When fully extended *Phascolosoma* may reach lengths of up to 15cm; when contracted they are only 4-5cm long.

Phascolosoma is very common in mussel beds and under rocks along semi-protected ares of the north coast. Generally when you find one, you will find many. There are other sipunculans which you may encounter in Humboldt Bay, but they are far less common than *Phascolosoma*.

PHYLUM ECHIURA **(serpent-like)**
the echiurans or spoon worms

Echiurans, like sipunculans, are unsegmented, coelomate worms; however, they do not possess tentacles around the mouth, have no introvert, and have the anus located in the more conventional posterior position. The most abundant echiuran along our coast is the distinctive *Urechis caupo* (previously mentioned as the "fat-innkeeper" or "weenie-worm"). *Urechis* is in many ways an atypical member of this phylum. Most echiurans possess a long flexible proboscis which bears grooves leading to the mouth; *Urechis* has but a very short folded proboscis with a simple groove at the front end. This animal resembles, among other things, a relatively large (up to 25cm) soft, pink sausage (fig. 44). The mouth is located at the base of the short grooved proboscis; the anus is at the opposite end and ringed by tiny gold-colored hooks or setae. *Urechis* lives in a U-shaped burrow in the mud flats. Water is pumped through the burrow by rhythmic circular contractions of the body forming narrow areas separated by thick bulges which pass from anterior to posterior along the worm. This is the same kind of muscular activity which forces food through digestive tracts (including yours) and is called **peristalsis**. The animal secretes a net of mucus through which the water flows. Periodically, the worm eats the trapped food, net and all, and then proceeds to construct a new sieve. As mentioned earlier, *Urechis* plays host to a variety of other animals which co-inhabit its burrow. These include *Hesperonoe adventor* (a scale worm), *Scleroplax granulata* (a small crab), and *Clevelandia ios* (fig. 145) (a tiny fish). Aside from gaining the protection of the burrow, the scale worm and crab also depend on the water current created by *Urechis* to supply their food. *Clevelandia* usually leaves the burrow to feed.

A small bit of information which can be employed to amaze your friends involves the method of gas exchange observed in *Urechis*. Rather than using gills or blood-filled tentacles for this purpose as do most "worms", this creature pumps water in and out of its anus, and oxygen and carbon dioxide are exchanged across the thin lining of the hind-gut (a form of breathing known as cloacal irrigation).

Reproduction in *Urechis* involves the males and females shedding their gametes into the water, external fertilization, and development through a trochophore larva to the young worm.

PHYLUM ARTHROPODA (jointed feet)

the insects, spiders, crabs, shrimps, etc.

Arthropods, arthropods, arthropods! Everywhere one looks there are arthropods. No other group of animals has so many species or so many individuals as the Phylum Arthropoda. They have invaded every conceivable habitat and dominate many. The most obvious example of abundant arthropods is the terrestrial insects. In the sea, however, it is the crustaceans (crabs, shrimps, lobsters, etc.) that have flourished and diversified and are our primary concern in this book. A few insects have returned to the sea to become part or full-time intertidal residents, and they, along with the so-called "sea spiders" (Pycnogonida) represent the only non-crustacean marine arthropods of significance.

Arthropods are obviously segmented but generally have the body highly regionalized beyond the wildest dreams of any heteronomous polychaete. Their appendages are jointed, a necessity brought about by their rigid outer covering or **exoskeleton**. While the mechanical and ecological advantages of this skeleton have certainly played major roles in the success of arthropods, being encased has also caused some difficulties. Movement despite the skeleton has been solved by joints in the legs and flexible areas between body segments. Growth is another problem altogether. The hard, non-living skeleton encases the animal in a suit of armor and restricts its size. The solution to this problem involves a complex, hormonally controlled loss of the skeleton (a process referred to as **molting** or **ecdysis**) and the concurrent formation of a new and larger one into which the animal can grow. The molting of a large crab is a dramatic process and may often be witnessed in the Dungeness crab in Humboldt Bay during the summer. The cast-off exoskeletons look just like the live crabs, and include the complete covering of the body, gills, the compound eyes (unique to arthropods), spines, even the hard lining of the stomach. Immediately following the molt, the animals are soft (hence the term soft-shelled crab for recently molted ones) and particularly vulnerable to injury and predation. They generally become secretive during this time, hiding in protected areas until the new skeleton is fully hardened. The jointed appendages and arrangement of body segments are matters of great importance in enabling one to trace the evolutionary relationships between various arthropod groups.

In addition to the exoskeleton and jointed legs, the arthropods display some other major changes from the conditions seen in their presumed ancestors, the Annelida. The coelom of arthropods is greatly reduced, and the main body cavity is an open (not enclosed by a mesodermal sac) blood-filled space (the **hemocoel**) in which the organs are bathed directly in the blood. Mating, reproductive biology, and development are extremely variable within the Arthropoda. Fertilization is generally internal, and some degree of parental care for the embryos is common. In some instances the embryos develop directly to miniature adults, in others (especially the marine forms) various larval stages may precede the juvenile form. The overall duration of larval life is generally related to the amount of yolk (nutrient) present in the egg. In cases of small quantities of yolk, the embryos must quickly become capable of feeding and thus they develop rapidly into larvae; where sufficient yolk is present, the embryos take longer to develop and there is a tendency towards direct development (no larvae), rather than free larval stages.

The present-day arthropods are generally divided into two major subphyla, the Chelicerata and Mandibulata. The former includes the Merostomata (horseshoe crabs of the eastern seaboard), the Arachnida (spiders and their kin), and usually the Pycnogonida (odd little creatures called the sea spiders, discussed below, which are not spiders at all). The mandibulates are the Insecta (the insects), the myriapods (centipedes, millipedes, etc.), and the Crustacea. The two subphylum names refer to differences in certain head appendages which, among other traits, serve to separate the two groups. The chelicerates have **chelicerae** as biting or grasping claw-like appendages on the front of the head, and no antennae; the mandibulates possess **mandibles** near the mouth for chewing, preceded by one (insects and myriapods) or two (crustaceans) pairs of sensory antennae on the head. Rest assured that the vast majority of marine arthropods which you encounter are crustaceans; the others are briefly summarized below.

Non-crustacean Intertidal Arthropods

A few types of bugs, beetles, and flies may be frequently encountered along the north coast. These air-breathing insects will most often be seen among piles of seaweed washed up on sandy beaches, or on algal covered rocks and mussel beds of the mid- to upper-intertidal zones when the tides is out. Most of these animals feed on dead algae and many lay their eggs in algal mats where the larvae develop and feed. In most cases their specific identification requires careful microscopic examination, and those of you who may be interested in such things are referred to L. Cheng's excellent treatise, **Marine Insects** (see reference list).

We have, on numerous occasions, seen a small red centipede inhabiting the vertical rock faces in the spray zone, originally pointed out to the northern author by the late Fred Telonicher, one of the finest naturalists to walk our beaches and a man whose knowledge of our shores was exceeded only by his love of nature and his fellow man.

While some biologists have spent a lifetime trying to unravel the systematics and biology of pycnogonids, we hope that the reader will be satisfied with the brief discussion that follows. Sea spiders may not even be chelicerates, but they are usually placed within that taxon for convenience. Intertidal pycnogonids are small—a dime-sized individual would be a fine catch (most are only a few mm across)—but some of the very-deep-water species may have leg spans of 80 or 90cm! *Pycnogonum stearnsi* (fig. 45) is relatively common all along the California coast, but several other species may be frequently

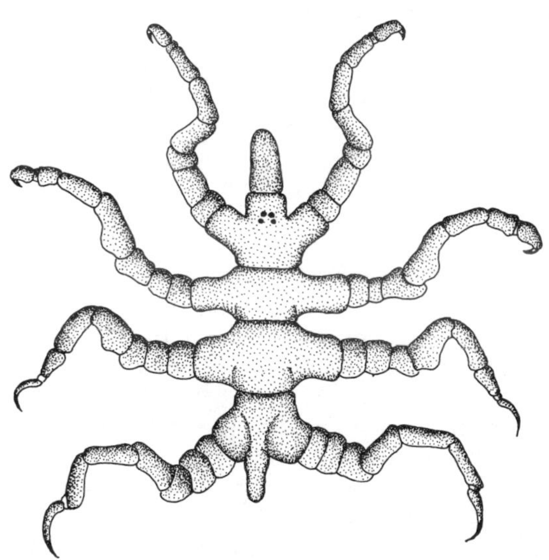

45. *Pycnogonum stearnsi* (~ 3mm long)

encountered if one knows where to look for them. Pycnogonids are most often found intertidally on hydroids or bryozoan colonies, and around the bases of some sea anemones. They feed by either piercing the animal's skin and sucking out its body fluids, or actually nipping off individuals from a colony and consuming them in their entirety. During the mating process the male sea spider fertilizes the eggs as they pass from the body of the female. The male then gathers up the fertilized eggs and carries them on a special pair of legs modified for brooding the developing embryos, which are eventually released as free-swimming larvae. If you find a pycnogonid carrying eggs, it is a male.

Crustaceans

Crustaceans occur in almost all sizes, shapes, and colors, and display habits of every imaginable type. Some are easily recognized as shrimps, crabs, barnacles, etc., while others, equally as important, are far less familiar to the amateur. The terminology of those who study crustaceans (carcinologists) is formidable, yet a smattering of it is necessary for purposes of description and identification. Reference to Figure 59 will help in locating the structures discussed below. Most larger crustaceans have their bodies divided into three regions, the head (**cephalon**), **thorax**, and **abdomen**, usually recognizable by bearing distinct types of appendages. Primitively each segment of each region bears a pair of appendages, although a great deal of variation exists within the group. The head consists of five fused segments and bears two pairs of sensory antennae, and three pairs of mouth parts (**mandibles**, and two sets of **maxillae**). The thorax includes a variety of possible appendage types from walking legs and claws, to some which may be modified as additional mouth parts. The adult thoracic legs are collectively termed **pereopods**. The head and part or all of the thorax may be covered dorsally and laterally by a shield-like **carapace**. The abdomen often bears appendages used for swimming (**swimmerets**) or other appendages, all called **pleopods**, and a terminal pair of **uropods**. The uropods, combined with the **telson** (a small, often platelike structure at the tip of the abdomen) may form a tail fan as seen in many shrimps and lobsters.

Only a few of the vast numbers of crustaceans are included in this guide, and many of the very tiny ones are treated only superficially or omitted completely. It is somewhat painful to treat the crustaceans in this manner since they occupy most of the authors' waking hours, yet to consider them in the depth they deserve would require several volumes. The classification presented in Appendix Two gives one some idea of the great diversity within this class.

Subclass Copepoda (the copepods)

Copepods are all quite small, the intertidal forms rarely exceeding 3mm in length. They are, however, extremely abundant, and the

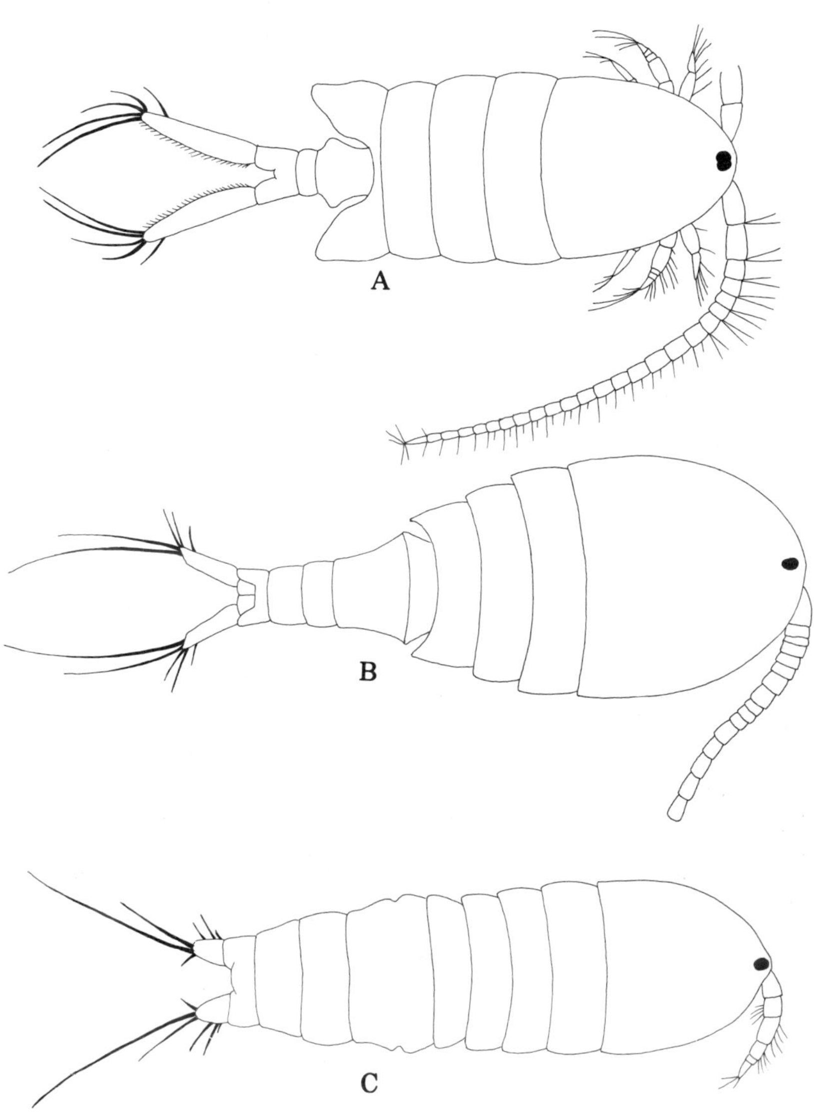

46. Copepods: A, calanoid; B, cyclopoid; C, harpacticoid (general form of bodies)

planktonic forms play a vital role as the primary consumers in the ocean's food chain. Most of these planktonic types are calanoid copepods, characterized by a pair of very long antennae used for swimming. Nearly all of the intertidal types are either harpacticoid or cyclopoid copepods. Figure 46 illustrates the three catagories. Copepods move about in short, jerky progression over the surfaces of algae and the undersides of rocks, or through the water. A hand lens will prove useful in seeing a bit of detail on these creatures. Most copepods retain the characteristic single larval or naupliar eye which appears as a distinct red spot in the middle of the head and lack the paired compound eyes of most arthropods. One of the most easily seen copepods is *Clausidium vancouverense*, a small red species that spends its life crawling along on the surface of the pink ghost shrimp *Callianassa californiensis* (fig. 60). Several groups of copepods are almost exclusively composed of commensal or parasitic members. Some of these may be seen on the skin of various local fishes, especially some of the small sharks in Humboldt Bay and similar habitats. These copepods attach loosely to the soft skin around the bases of the fins or near the gill openings.

Subclass Cirripedia (the barnacles)

Adult barnacles superficially bear little resemblance to other crustaceans, and until 1830 were included in the Phylum Mollusca. The barnacles are sessile crustaceans, firmly cemented to hard substrates, and covered with numerous calcareous plates (except for a few odd parasitic ones). The free-living ones live within the shells with their heads directed down towards the substrate. The highly modified thoracic legs are used to filter food from the surrounding water or prey upon small animals which happen within range. The legs may rhythmically sweep through the water or be held extended for several minutes before being pulled inside. Once the legs are retracted, the collected food particles are carried to the mouth and swallowed. This feeding activity may be observed when the animals are submerged, but they close up tightly when exposed at low tide. Barnacles are one of the few crustacean groups whose members are hermaphroditic, that is each individual contains both male and female reproductive organs. The male system includes a dramatically long penis capable of reaching out to impregnate neighboring barnacles. Mating may be witnessed when the animals are under water during the breeding season (mostly spring and summer). The embryos are brooded a short while, then released into the water as larvae, which eventually settle and attach.

The acorn or sessile barnacles (Balanomorpha) are characterized by having their shells cemented directly to the substrate (not elevated on a stalk). *Balanus glandula* (fig. 47) is among the most common of these volcano-shaped barnacles and may be found all along the Pacific coast on rocks or pilings in the high tide zone where it spends much of its time exposed to air. By being capable of surviving these rigorous conditions, it avoids competition for space with numerous other

47. *Balanus glandula*
(~ 1cm diameter)

48. *Balanus cariosus*
(~ 2cm diameter)

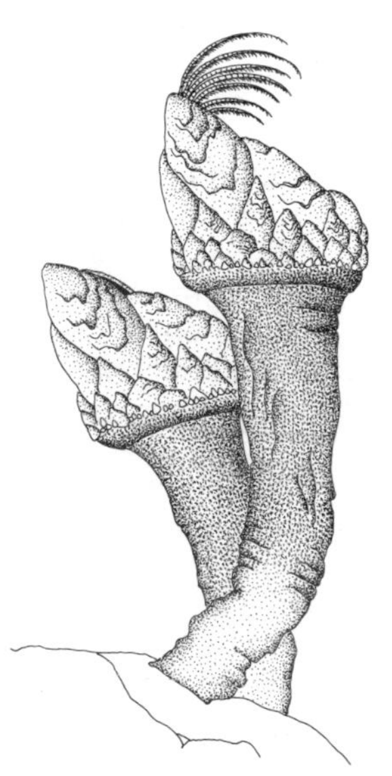

49. *Pollicipes polymerus*
(~ 5cm high)

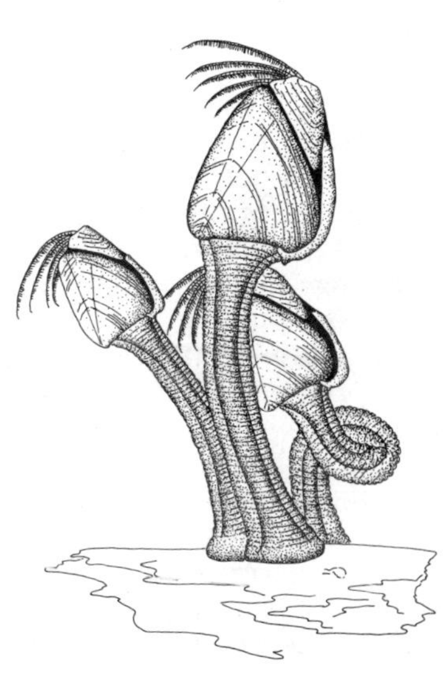

50. *Lepas anatifera*
(~ 4cm high)

organisms less tolerant of exposure and which must therefore live in lower zones. This barnacle has been the object of a number of intensive studies by west-coast marine biologists, some of which have produced insights into fundamental concepts of ecology. Another common barnacle is *Balanus cariosus* (fig. 48). This species often occurs in high numbers in mussel beds. The plates of *B. cariosus* are heavily ribbed as opposed to the relatively smooth shells of *B. glandula*. In many areas competition for space results in such crowding that *B. cariosus* populations appear more as clusters of elongate, angular, white tubes than the typical volcano shape. *Chthamalus dalli* is a northern species of small barnacle, usually less than 5mm in basal diameter. *Chthamalus* is often confused with small specimens of *B. glandula* but can be identified by its rather low, somewhat rounded profile, and greyish rather than white color. Some species of acorn barnacles, despite their sessile life styles, are very efficient competitors capable of displacing (by crushing or undercutting as they grow) other sessile invertebrates and algae.

The goose or stalked barnacles (Lepadomorpha) are characterized by having their shell-encased body supported by a fleshy or leathery stalk. The only true intertidal lepadomorph is *Pollicipes polymerus*, the very common goose barnacle (fig. 49). The stalk supports the animal above the substrate (and above some of its competitors), allowing the individual to twist more than acorn barnacles to "face" the flow of water for more effective filter feeding, and adds flexibility in areas of high wave action. *Pollicipes polymerus* and the California mussel, *Mytilus californianus* (fig. 110) are two of the predominant sessile invertebrates of Pacific Northwest seashores. Their abundance has prompted several workers to investigate their individual ecology and interaction (they often occur together). *Mytilus*, it appears, is competitively dominant to *Pollicipes* and, given time, will replace and crowd out the latter. This sequence of events may be prevented from reaching its conclusion (a "pure" stand of *Mytilus*) by such factors as drifting logs that crash into mussel beds, or predation by seastars (see Ecology section).

Lepas anatifera (fig. 50) is another stalked barnacle which may be encountered attached to objects which have washed ashore during storms. This species is fully pelagic and occurs only on floating materials in the open ocean, such as driftwood, pumice, glass and cork fishing floats, and occasionally on the skin of sea turtles and whales.

Goose barnacles are so named because of a very old fable in which geese were described as being spawned from these marine animals. They were apparently so-named because their feeding legs resemble feathers. The story is usually attributed to a 16th-century writer known for his vivid exaggerations, John Gerard. The name "goose-neck" barnacle has no proper foundation in the literature and appears to be a misnomer of the original "goose" barnacle.

Subclass Malacostraca

The bulk of the familiar crustaceans such as shrimps, crabs, sand "fleas", etc., are members of the Malacostraca. The diversity of this group is great, and, aside from obvious forms, the reader will have to depend upon gaining a little experience in order to recognize many members of this taxon. With a little patience, however, the persistent tidepooler will eventually be able to spout names with a minimum of difficulty. Reference to the classification scheme presented in the appendix will help the reader make some order out of the following accounts.

The Malacostraca may be divided into two subcategories or "series". One is the Series Phyllocarida which includes the order Nebaliacea, represented locally by *Nebalia pugettensis*, which periodically occurs in high numbers in shallow water of Humboldt Bay. The nebaliids are characterized by having a bivalved carapace appearing as a pair of clam-like shells covering the thoracic region of the body. The unhinged sides of the carapace open ventrally where the animals create a feeding current from which food is strained by tiny hair-like setae. Nebaliids are generally less than 1cm long and a hand lens is required to see any detail.

The second category within the malacostracans is the Series Eumalacostraca. The animals in this group tend to have a fixed and consistent number of body segments and related appendages. As reference points, we can consider a "typical" case as a condition in which there are five head segments (usually fused, bearing successively from front to back the two pairs of antennae, mandibles, and two pairs of maxillae), eight thoracic segments (with eight pairs of limbs), and six abdominal segments (with five pairs of pleopods and one pair of uropods). Variations upon this basic arrangement, the nature of the carapace, and the structure of the compound eyes, will serve to characterize the various subgroupings of the Eumalacostraca included here.

Superorder Peracarida: The female members of this superorder are distinctive in the possession of large flaps called **oostegites** on the underside of the thorax; these flaps function as a brood pouch or marsupium. The embryos are retained beneath these plates during development until they hatch as miniature adults. Many peracarids lack a carapace altogether, and when it is present it is generally reduced. Of the six orders that are placed within the peracarids, only three are likely to be encountered by the casual visitor to our local shores.

The order Mysidacea includes shrimp-like animals called the opossum shrimps because of their habit of brooding the young. Near-shore mysids rarely exceed 2cm in length, although some oceanic types may be 15cm long. These filter-feeding crustaceans possess a well-developed carapace covering the thorax, but it is not fused to the last few segments. The presence of the carapace and the fact that their eyes are stalked easily separates these animals from the other peracarids discussed below (amphipods and isopods). In fact, mysids are more easily confused with true shrimps (Decapoda) than with the members of their own superorder, but they may be distinguished by carefully examining the carapace. The carapace of a decapod is fused dorsally with the entire thorax, whereas in the mysids, the carapace is loose over the last few thoracic segments. The most common local species of mysid is *Archaeomysis maculata* (fig. 51), although other types occur, especially to the south. The casual observer will easily overlook these nearly transparent animals, and sometimes only their flitting shadows can be seen. They are most abundant in shallow water over sandy bottoms. A few sweeps of a dip net in such an area will usually capture some mysids which may then be observed more closely in a bucket or jar before releasing them. The real challenge is catching them by hand!

Among the most abundant crustaceans in the sea are members of the order Amphipoda; it would be rare to sample any portion of the intertidal zone without collecting at least a few of these animals. Amphipods are characterized by the absence of a carapace and by having eyes which are flush with the sides of the head (a condition referred to as sessile eyes) rather than being raised on stalks. They have seven pairs of thoracic legs rather than the characteristic eight, the first pair having been incorporated into the head region as feeding structures called **maxillipeds**. The first two pairs of legs which persist on the thorax are often clawed or prehensile (modified for grasping) and differ markedly from the remaining five pairs. There are three principal suborders of amphipods, two of which, the Caprellidea and Gammaridea, are common intertidally. The third group (Hyperiidea) is made up of members which are exclusively planktonic. A few hyperiids, notably the genera *Hyperia* and *Hyperoche*, have been mentioned earlier as being found attached to or imbedded in medusae and ctenophores. The caprellids or "skeleton shrimp" are common on hydroids, bryozoans, bushy algae, and similar substrates in protected areas such as Humboldt Bay (fig. 52). These odd little amphipods more closely resemble walking sticks than crustaceans and often match their surroundings to a remarkable degree. They are generally small, less than 1cm in length. Most caprellids may be called lurking or ambush predators. They bob back and forth or pose motionless, attached to the hydroid by their rear legs. Using their ability to sense vibrations in the water they grasp small creatures which pass within reach of their large prehensile claws. They may also be seen to "inch" along the surface on which they live, picking at detritus and other microscopic bits of food in their environment.

By far the most numerous and common amphipods belong to the suborder Gammaridea. Most are small, usually less than 1cm long, although a few may exceed three times that length. Gammarids may be found throughout the intertidal region, especially among bushy algae on rocks. They appear as tiny shrimp-like creatures of a variety of colors, often crawling along on their sides by bending and straightening or swimming in circles. Most gammarids are markedly flattened from side to side (laterally flattened or **compressed**). This

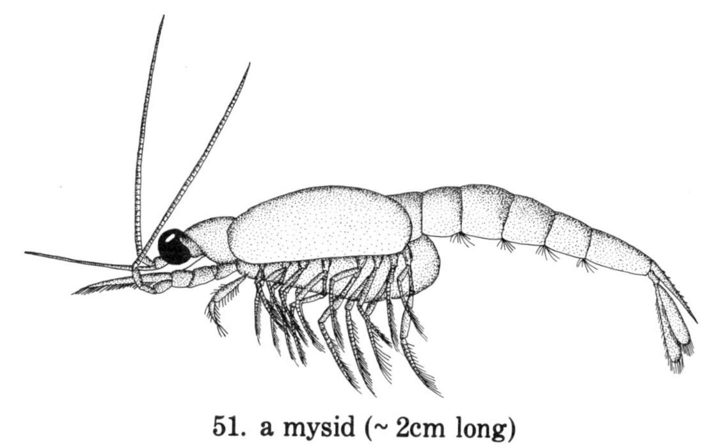

51. a mysid (~ 2cm long)

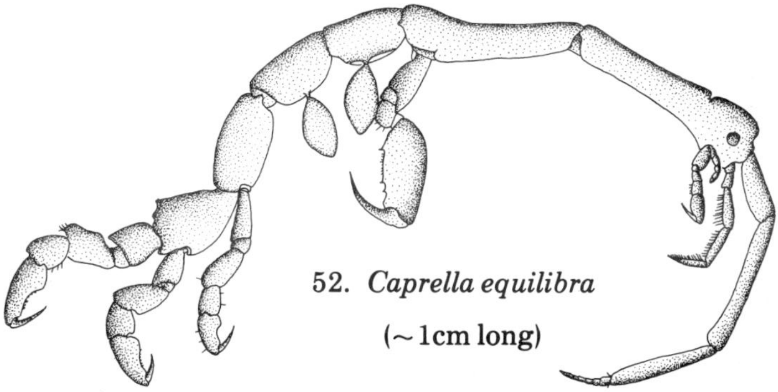

52. *Caprella equilibra*

(~ 1cm long)

53. *Orchestoidea californiana* (~ 2cm long)

general shape distinguishes them from most isopods (see below) which are flattened from top to bottom (dorsoventrally flattened or **depressed**). Most gammarids are extremely difficult for the novice to identify; their specific differences are often small parts of other small parts. That is not to say, however, that once you have seen one gammarid you have seen them all. Two genera may be easily located and serve to introduce the new beachcomber to the amphipods. *Orchestia* and *Orchestoidea* (fig. 53) are small and large beachhoppers respectively. These animals are high-tide zone inhabitants of sand beaches. They avoid direct sunlight and drying by burrowing a few centimeters into the sand, or secluding themselves under piles of washed-up algae and driftwood. In the daytime you may expose them by digging in the sand (their presence being indicated by patches of closely spaced, small holes), or turning over piles of algae. The beachhoppers are best observed, however, early on foggy mornings, or at night, when they leave their hiding places to scavenge along the sand surface. The distinct grey-green bodies and reddish antennae of *Orchestoidea californiana*, coupled with the ability of the animals to perform spectacular jumps and mid-air somersaults, make them the mini-clowns of the sandy beach scene. Most gammarids scavenge on animal and plant material. A few construct tubes in which they reside and filter the passing water for food. Because of their overwhelming abundance, the gammarids form an integral link in the food webs of most shoreline and shallow subtidal communities, providing a major food source for a host of other animals.

While the northern author finds his particular niche among a small group of amphipods, the southern author has seen fit to become fanatically dedicated to members of the order Isopoda. Neither of us understands the other's odd leanings but sigh, (in unison), "To each his own." The Order Isopoda includes the familiar sow bugs and pill bugs common in backyards and flower gardens. Marine forms are variously called sea slaters, rock lice, gribbles, etc. They may be distinguished from their close relatives, the amphipods, by being flattened top to bottom (depressed) and generally having all seven pairs of thoracic legs more or less alike. In addition, the isopods employ their abdominal appendages (pleopods) as gills, whereas the amphipods bear gills on their pereopods. Isopods are important scavengers on the world's seashores and, along with certain amphipods, play a major role in keeping beaches clean and recycling organic matter.

The isopods are a diverse group having invaded a number of unlikely habitats. They include the only truly successful terrestrial crustaceans. Some are parasites on fishes, others on shrimps, and still other destroy man's encroaching coastal structures by boring into docks, floats, and other unnatural fabrications. Many isopods are easily identified, at least to the level of family or genus. The key below will help in recognizing some of these more distinctive and common forms, including two terrestrial species likely to be encountered near the seashore.

Key to Some Common Types of North Coast Isopods

1. Terrestrial or on high spray-zone rocks. 2
1. Mid- to low-intertidal (or subtidal), on various substrates. 4
2. Relatively large, usually over 2cm long, mottled grey isopods; restricted to spray zone. . . . . *Ligia* (figs. 54-55)
2. Never more than 2cm long; fully terrestrial. 3
3. Cannot roll into a ball. . . . . *Porcellio scaber*
3. Capable of rolling into a ball. . . . . *Armadilloniscus lindhali*
4. Parasitic on fishes or crustaceans. 5
4. Free-living or boring in wood. 6
5. Commonly attached under the side of the carapace of bay shrimp (*Crangon*). . . . . family Bopyridae
5. Attached to the gills or in the mouth cavity of fishes. . . . . family Cymothoidae
6. Boring in wood (minute isopods, usually less than 4mm long; produce extensive and noticeable burrows in wood which is submerged in sea water). . . . . *Limnoria*
6. Not boring in wood (although some may be found inhabiting the burrows of other animals in "rotting" timbers). 7
7. Uropods attached laterally and visible as a "tail fan" when viewed dorsally (fig. 56). 8
7. Uropods folded beneath the abdomen as two "barn doors" covering the abdominal gills. . . . . family Idoteidae (fig. 57)
8. Usually capable of rolling into a ball . . . . . . family Sphaeromatidae
8. Incapable of rolling into a ball. . . . . family Cirolanidae (fig. 56)

Two species of true terrestrial or land-dwelling isopods may be found under boards or loose stones near the coast (as well as inland). They generally forage as scavengers during the night. Their abdominal appendages, or pleopods, have been converted from gills to air-breathing structures. Close inspection of the underside of the abdomen will reveal several patches of whitish tissue on some of the pleopods where gases are exchanged between the blood of the animal and the environment. The common grey to black sow bug is *Porcellio scaber*. Its less common relative which can roll into a ball is *Armadilloniscus landhali*.

Members of the genus *Ligia* are restricted to the spray zone, although many will be seen to follow a receding tide part way down the beach. The most common species along the extreme north coast of California are *Ligia pallasii* (fig. 54) and *L. occidentalis* (fig. 55). The latter is more abundant further south. Although *Ligia* can tolerate periods of submergence in sea water for over 48 hours, it prefers the spray-zone habitat. It cannot, however, withstand low humidity or other drying conditions, and must keep its abdominal gills moist at all times. These animals may occasionally be seen dipping their forked uropods into small pools of sea water to draw moisture onto the gill surfaces. Members of this genus are scavengers on stranded materials on the beach, and they are instrumental in the recycling of both animal

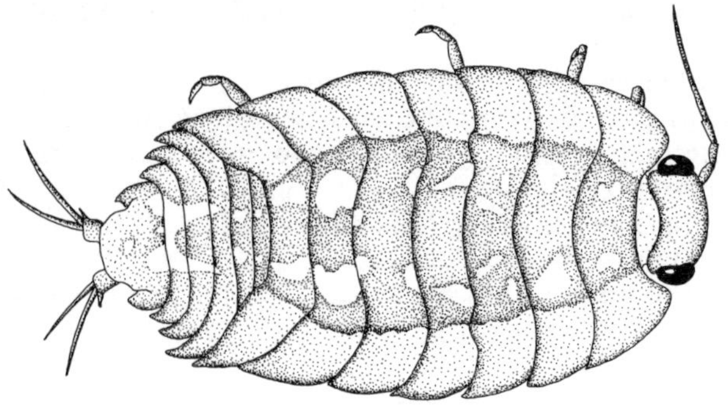

54. *Ligia pallasii* (~ 3cm long)

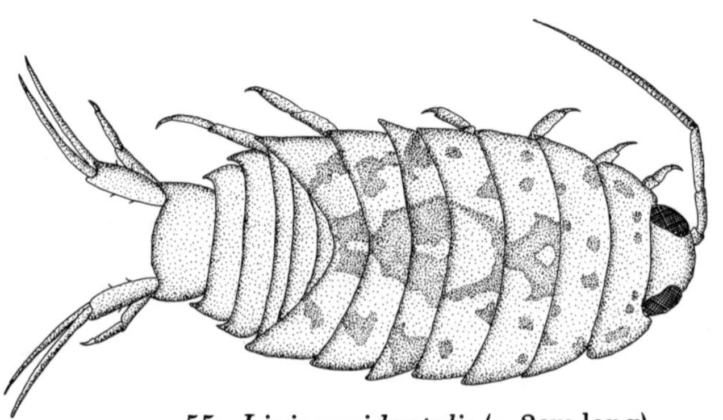

55. *Ligia occidentalis* (~ 2cm long)

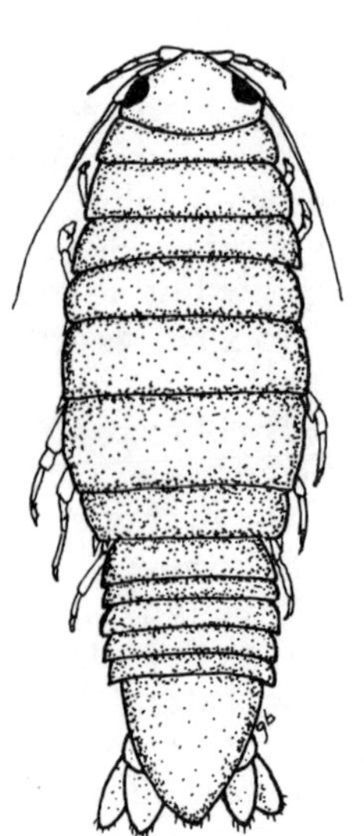

56. A cirolanid isopod (~ 1cm long)

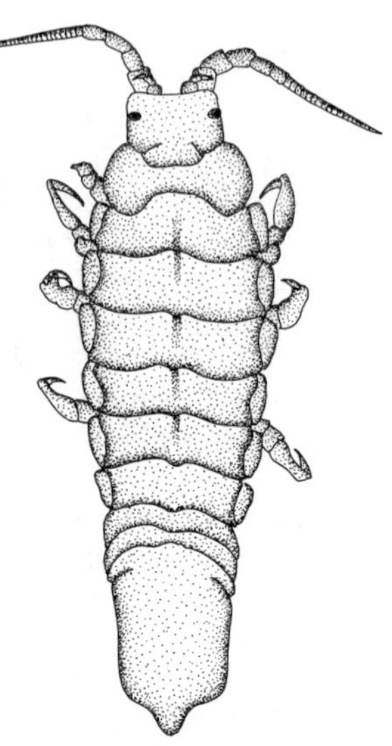

57. *Idotea stenops* (~ 4cm long)

and plant material cast upon the shore. They may be seen out and about on early foggy mornings, or at night, scurrying over the rocks and hard-packed sand in search of food left by the receding tide. At other times they take refuge from the sun by remaining in the damp caves along the shoreline cliffs (*L. pallasii*) or in cracks in high rocks nearer the sea (*L. occidentalis*).

Parasitic members of the families Bopyridae and Cymothoidae are common along our coast but generally go unnoticed by the casual observer. One of the bopyrids lives under the carapace of a local shrimp of the genus *Crangon* (see Decapoda), and may be located by looking for a large hump on one side of the host. The large female isopod may be seen by lifting the edge of the shrimp's carapace. The male is generally much smaller and far less obvious. The cymothoids, or fish lice, are common parasites on local rock and bottom fishes. These animals suck blood and cause tissue damage to the host. The two predominant species on the north coast are *Lironeca californica* and *L. vulgaris*. Checking the gills and mouth cavities of fishes taken by hook and line will often turn up a few of these animals.

The weakening and destruction of submerged timbers and pilings are in part, due to the remarkable habits of the isopod *Limnoria*, the gribble. The effect of these borers is most easily observed on old pilings at low tide. Many will appear to have been gnawed away by beavers when viewed from a distance, but closer examination will reveal that the wood has been riddled with tunnels. These animals live within a definite zone so that the effects of their feeding give older pilings an hourglass shape. *Limnoria* may be found by breaking off a piece of such wood and searching for these small whitish sow-bug-like isopods. These and other boring animals cause incredible amounts of damage to man-made coastal structures, but then, they were there first.

The majority of relatively large and obvious free-living isopods fall into three families (Sphaeromatidae, Cirolanidae, and Idoteidae). There are, of course, many other types, but many are small and not seen without the aid of a hand lens, or do not occur intertidally on our coast. The sphaeromatids and cirolanids are typically oval, pill-bug-shaped isopods, commonly found in sandy areas, in mussel beds, or on the undersides of rocks (fig. 56). *Excirolana linguifrons* is extremely abundant buried near the mid-tide levels in wave-swept, sandy beaches such as the area along the Samoa Peninsula. The idoteids are more elongate, green, brown, or reddish, and usually attach to marine plants. Some members of the genus *Idotea* exhibit remarkable cryptic coloration, and often match precisely, both in color and shade, the plant to which they are attached. The largest local intertidal isopod is *Idotea stenops* (fig. 57) which may reach a length of 5 or 6 centimeters. This species is almost always found aligned on the long axis of the main stipe of the feather-boa kelp, *Egregia menziesii* (fig. 163). The camouflage is so accurate that this isopod is most easily found by running one's fingers along the kelp and feeling for the animal. Other species of idoteids attach in similar manners to other algae, as well as to the surf grass *Phyllospadix*. The strong desire to cling to a suitable substrate can easily be observed by placing one of these isopods in a container, and putting various objects in with it. Any elongate, narrow artifact will substitute for the plant on which they normally dwell, and their attaching behavior may be quickly observed. You will note that the thoracic legs are equipped with hook-like terminal segments well suited for grasping.

Superorder Eucarida: This group indludes, among other things, the shrimps, lobsters, and crabs. The eucarids are united by thier possession of a well-developed carapace which covers, and is dorsally fused with, the entire thorax. Of the two orders, only the Decapoda is well represented in the intertidal zones. The Order Euphausiacea is a group of planktonic, shrimp-like crustaceans, three or four centimeters long, commonly known as "krill". Their abundance in certain regions makes them very important food items in the diets of many fishes, sea birds, and some whales.

The decapods are characterized by having five pairs (usually) of well-developed walking or grasping legs on the thorax; hence the name Decapoda, ten feet. Recalling the "standard" eight pairs of thoracic legs in the malacostraca, we find that the first three have become mouth parts (maxillipeds) in the decapods. The hermit crabs and their kin (anomurans) have the last pair or two of thoracic legs reduced, leaving only three or four pairs readily visible. The following key will help in the identification of many local decapods, but do not be surprised to find a few others which do not fit the descriptions.

Key to Some Common North Coast Decapods

1\. Body elongate, shrimp-like (shrimps) (figs. 58, 59, 60)........... 2

1\. Body not elongate (crabs, hermit crabs, mole crabs)............ 3

2\. (Note three choices) Large (>10cm), pinkish or blue-grey "shrimps"; burrowing in mud flats... *Callianassa* (pink) (fig. 60) & *Upogebia pugettensis* (blue-grey)

2\. Brownish-grey, with black markings, 5 or 6cm long; body more or less straight, without an obvious angular bend; not greatly compressed; on sandy bottoms or among eelgrass in bays............ *Crangon* (fig. 58)

2\. Usually in tidal pools or on eelgrass; often distinctly and variably colored (red, green, striped) or clear; body with a distinct angular bend in abdomen; obviously compressed.......... *Heptacarpus* & *Hippolyte* (fig. 59)

3\. Living in snail shells (hermit crabs) (fig. 64).................... 4

3\. Not living in snail shells... 6

4\. Legs banded with blue or white 5

4\. Legs not banded with blue or white .... *P. granosimanus*

5\. Antennae reddish, different from color of body ....*P. samuelis*

5\. Antennae not reddish, nearly same color as body .... *P. hirsutiusculus*

6\. Oval, grey; smooth and streamlined; burrowing in wave-swept sandy beaches (mole or sand crabs)...... *Emerita analoga* (fig. 65)
6\. Body more or less crab-like; not found burrowing in wave-swept sandy beaches.. 7
7\. Small pale crabs living within clams.................. *Pinnixa*
7\. Not inside clams.. 8
8\. Small crabs, rarely exceeding 1cm across carapace, commensal in the burrows of *Urechis* and *Callianassa* in bay mud and sand flats.................................. *Scleroplax granulata*
8\. Not found by excavating mud flats; if partially buried in mud the crabs are large, at least several centimeters across the carapace.. 9
9\. Carapace expanded laterally to cover legs..................... .... *Cryptolithodes sitchensis* (fig. 61)
9\. Carapace "normal", legs usually visible when viewed dorsally.. 10
10\. Claws followed by three pairs of obvious walking legs; large antennae arise outside (lateral to) the eyes.......................... 11
10\. Claws followed by four pairs of walking legs; antennae arise between (medial to) the eyes....................................... 12
11\. Claws "hairy"........................... *Pachycheles* (fig. 63)
11\. Claws relatively smooth.................. *Petrolisthes* (fig. 62)
12\. Front of carapace produced forward (as a rostrum) well beyond reach of eyes.. 13
12\. Front of carapace not greatly produced beyond reach of eyes... 14
13\. Carapace smooth, dark green or brown. *Pugettia producta* (fig. 67)
13\. Carapace "hairy" and rough..... *Loxorhynchus crispatus* (fig. 66)
14\. Carapace more or less square.................................. 15
14\. Carapace oval, distinctly wider than long..................... 17
15\. Carapace greenish with distinct reddish transverse lines (rocky intertidal)...................... *Pachygrapsus crassipes* (fig. 70)
15\. Carapace without reddish transverse lines.................... 16
16\. Carapace reddish; legs smooth; with red spots on claws (rocky intertidal)......................... *Hemigrapsus nudus* (fig. 71)
16\. Carapace greenish; legs hairy without red spots on claws (restricted to bays)........................ *Hemigrapsus oregonensis*
17\. Tips of claws black.. 18
17\. Tips of claws not black................. *Cancer magister* (fig. 69)
18\. Teeth along sides of carapace -shaped; with numerous red spots on underside of body................ *Cancer antennarius* (fig. 68)
18\. Teeth along sides of carapace -shaped; without numerous red spots on underside of body.................... *Cancer productus*

The "true" or swimming shrimps belong to the suborder Natantia, which includes many of the offshore commercially important shrimps. Several of these natant decapods may be found intertidally. Species of the genus *Crangon* (fig. 58) occur in relatively high numbers in Humboldt Bay. Some of these are taken by commercial dredge boats operating in deeper waters from San Francisco to the Canadian border. These shrimps generally occur on sandy bottoms and display remarkable protective coloration patterns. At least two species may be

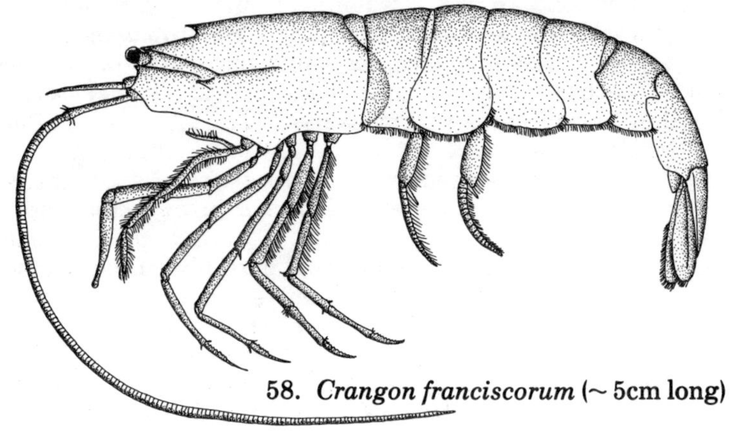

58. *Crangon franciscorum* (~ 5cm long)

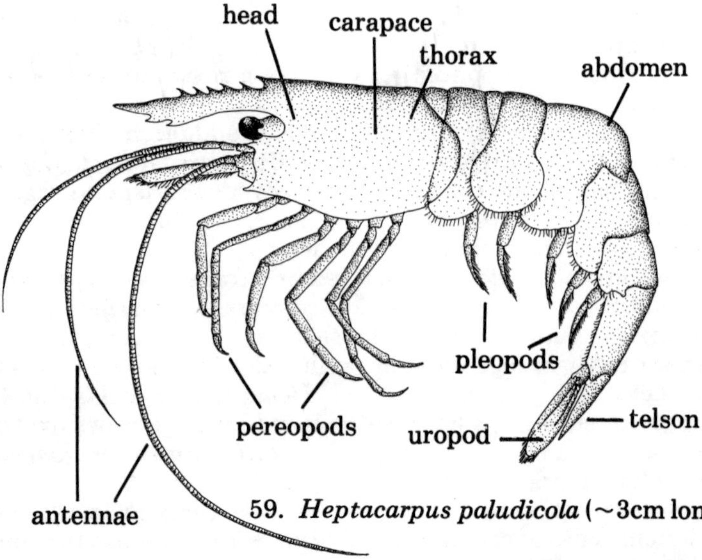

head
carapace
thorax
abdomen
pleopods
pereopods
uropod
telson
antennae

59. *Heptacarpus paludicola* (~3cm long)

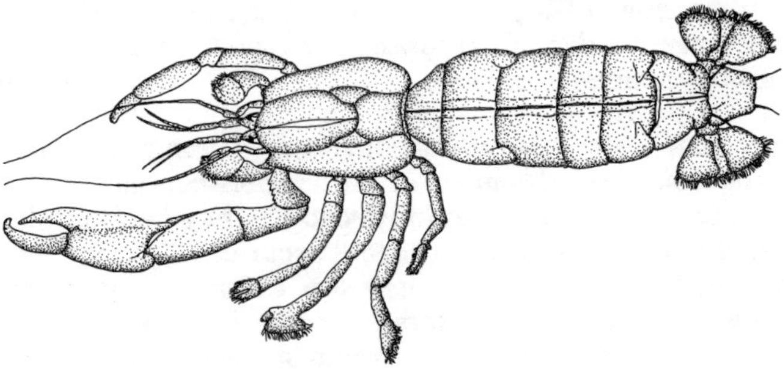

60. *Callianassa californiensis* (~ 10cm long)

found in Humboldt Bay, *Crangon nigricauda* and *C. franciscorum*. Both are bottom-feeding scavengers on plant and animal matter. *Hippolyte californiensis* is a bright grass-green shrimp measuring around 3cm in length. You will most commonly encounter this species among blades of eelgrass, which they closely match in color, in bays. Members of the genus *Heptacarpus* (fig. 59), close relatives of *Hippolyte*, are more common in tidal pools and may occur in a variety of colors, including green, red, brown, and sometimes striped and often match their background.

Members of the suborder Reptantia are decapods which are, for the most part, bottom dwellers and adapted for walking rather than swimming. The classification of certain higher taxa within the Reptantia is still being debated among the specialists. The scheme we have chosen is but one of several possibilities. The Section Anomura includes ghost shrimps, hermit crabs, porcelain crabs, and mole crabs, each type belonging to a different family. The family Callianassidae contains the pink ghost shrimps, *Callianassa californiensis* (fig.60) and the larger *C. gigas*, common inhabitants of local muddy tidal flats. The locations of the complicated burrow systems of these animals may be identified by volcano-like mounds on the surface of the substrate. These large "shrimps" have rather soft bodies and require the protection of their burrows and the physical support of the surrounding water. The single large formidable-looking claw is primarily a digging tool and actually quite ineffective as a weapon. This claw is longer and more slender in *C. gigas* than in *C. californiensis*. These animals draw water through their burrows by fanning the flap-like abdominal appedages and filter-feed upon small animal life by straining the water with fine hair-like spines. Another similar animal may be encountered in Humboldt Bay. The blue mud shrimp, *Upogebia pugettensis*, may be recognized by its faint bluish coloration and the lack of an oversized claw.

The porcelain and umbrella crabs belong to the families Porcellanidae and Lithodidae respectively. All of these anomurans closely resemble the true crabs (brachyurans, see below) but may be distinguished from them by the four, rather than five, well-developed pairs of thoracic legs. Also, the antennae of anomurans arise lateral to their eyes, but in brachyurans they arise between the eyes. The umbrella crab *Cryptolithodes sitchensis* (fig. 61) is occasionally seen in the low intertidal rocky areas of the north coast. The tent-like carapace is extended so far to the sides that it renders the legs invisible when viewed from the top, and this animal is frequently overlooked or passed by as an uninteresting lump or rock on the bottom. Most of the local *Cryptolithodes* are red and measure four or five centimeters across the carapace. *Petrolisthes* (fig. 62) includes the smooth porcelain crabs. Their upper surface is generally a deep reddish-brown, and they rarely exceed 15mm in width of the carapace. The flattened body and claws adapt these animals for an under-rock life in the mid-intertidal areas of semiprotected shores. Unlike almost all other crabs, these creatures obtain their food by filter-feeding rather than

UMBRELLA CRAB

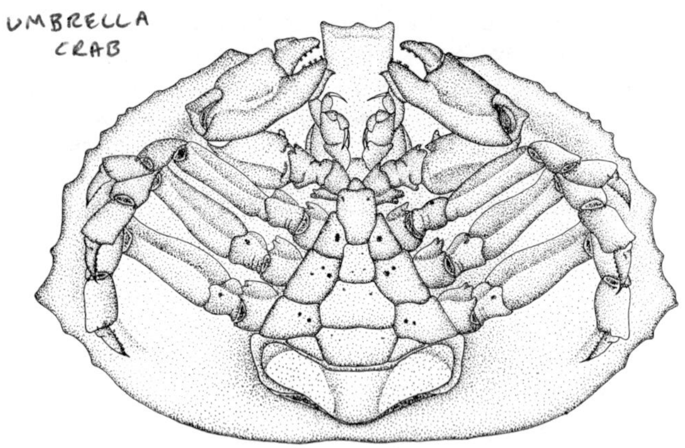

61. *Cryptolithodes sitchensis*, ventral view (~ 4cm wide)

"SMOOTH" PORCELAIN CRAB

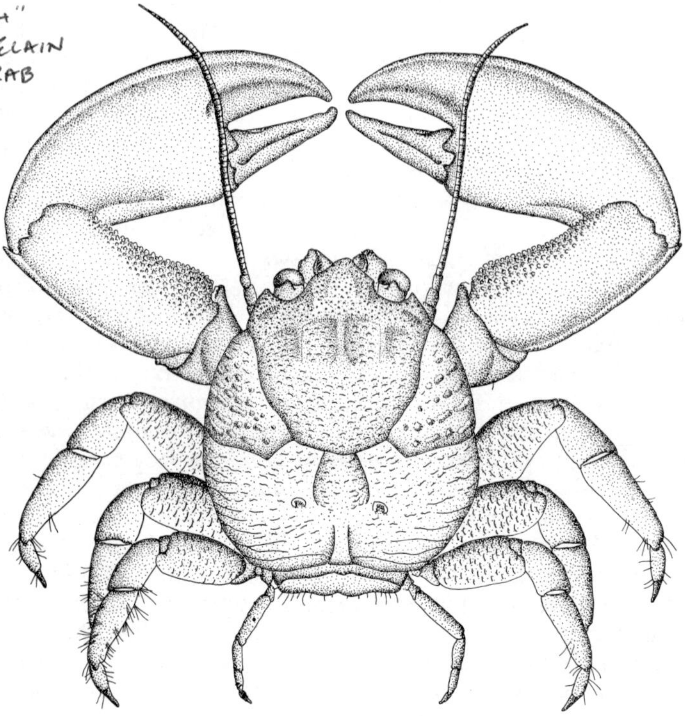

62. *Petrolisthes cinctipes* (carapace ~1.5cm wide)

scavenging or predation. Water currents are created by the beating of mouth parts, and comb-like edges filter food. The common local species are *Petrolisthes cinctipes* and *P. eriomerus*. The "rough" or "hairy" porcelain crabs along our coast are of the genus *Pachycheles* (fig. 63). They are similar in size and general shape to *Petrolisthes*, but have numerous bumps and "hairs" on the claws. Their range of habitats includes pilings and algal holdfasts as well as the undersides of rocks. They are generally not encountered with the frequency of *Petrolisthes* along our shores.

The most familiar anomurans are the hermit crabs (Family Paguridae) which characteristically take up residence in the empty shells of snails. Their abdomens are soft, lacking a hard exoskeleton, and the borrowed shells afford protection from damage and predation. Hermit crabs are important scavengers in intertidal zones around the world, scampering about the rocky pools and tidal flats, picking up bits of organic debris and generally helping to keep the place tidy. Some species have also been observed feeding on living algae or exhibiting a behavior that suggests they may occasionally filter-feed in a manner reminiscent of the porcelain crabs. The common intertidal hermit crabs of our coast belong to the genus *Pagurus* (fig. 64). The key separates three local species. These crabs have the last two pairs of thoracic legs much reduced and only three pairs, including the claws, protrude from the shell as the animal moves. The soft abdomen is twisted to fit the inner coil of the shell. A hermit crab may be removed from its home in order to see its hidden features and to observe re-entry behavior first hand. Hold the shell tightly, without jiggling it, and suspend the crab a centimeter or two above the water's surface. When the animal "realizes" that its shell is held fast, it will stretch out and touching the water, eventually drop free. Not all individuals will be cooperative, and any rough treatment will cause the animal to quickly retreat inside its house. Be sure to allow the crab to return to its shell, as hermit crabs are extremely vulnerable when exposed, and it is not always easy for them to find empty shells.

Among the strangest and most specialized anomurans are the so-called mole or sand crabs (family Hippidae). *Emerita analoga* (fig. 65) is the local species, occuring on wave-swept beaches such as the Samoa Peninsula, although they are much more common to the south. Small individuals of less than about 10mm are males, while those larger than about 15mm are females. The reason for the size difference is not gluttony by the fair sex, but that the young animals are males and undergo a sex reversal later in life when they are larger to become females. At least this is the standard explanation, although we understand someone has recently stirred the pot on this question, and we await more information. This sort of male-to-female sex reversal is call **protandry** and is not an uncommon phenomenon among a variety of invertebrate groups. Surgically conducted activities in humans which produce similar end results cannot, however, be considered true protandry. *Emerita* is an oval greyish crab which burrows into the sand along the upper mid-tide level. During low tides they may be

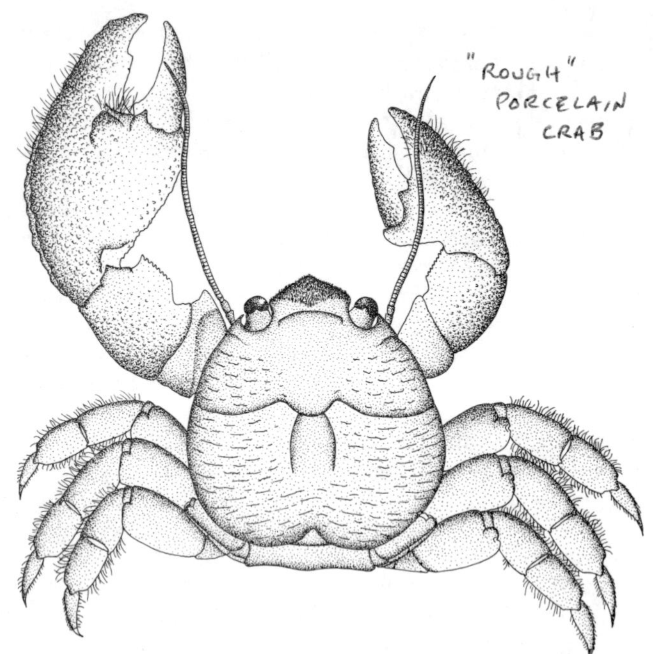

63. *Pachycheles rudis* (carapace ~ 2cm wide)

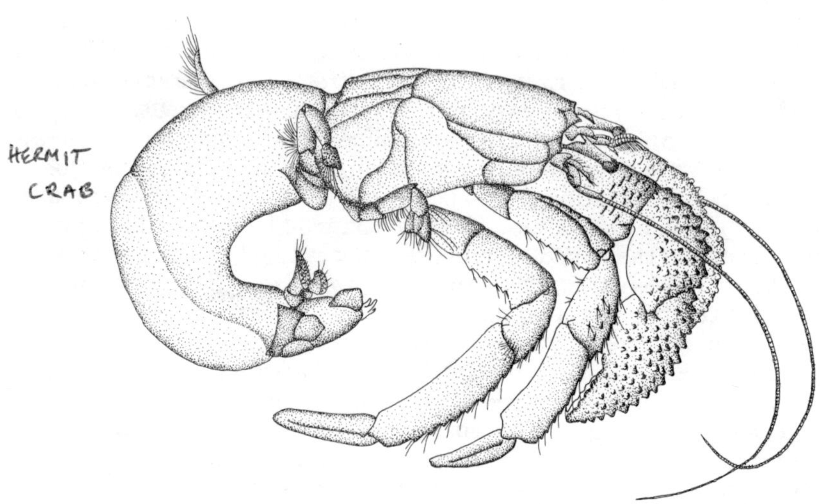

64. *Pagurus hirsutiusculus* (~ 3cm long), removed from shell.

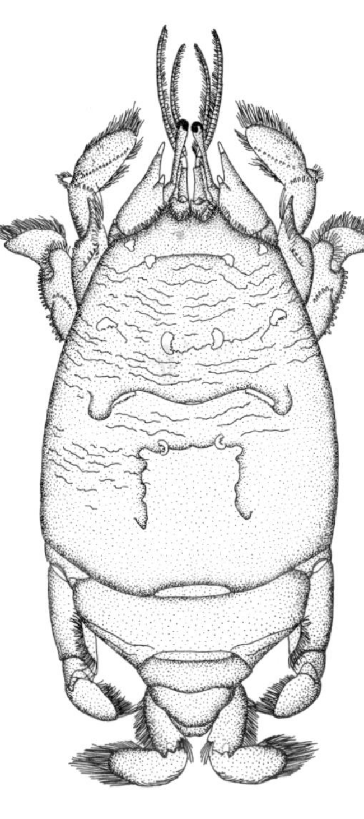

65. *Emerita analoga* (~ 3cm long)

SAND OR MOLE CRABS

SPIDER OR DECORATOR CRAB

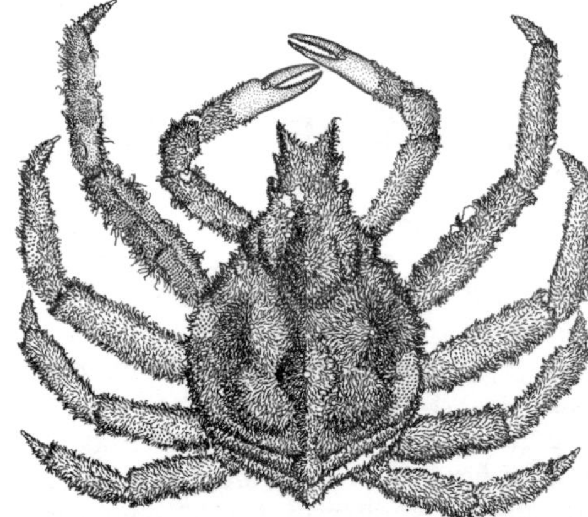

66. *Loxorhynchus crispatus* (carapace ~ 2cm wide)

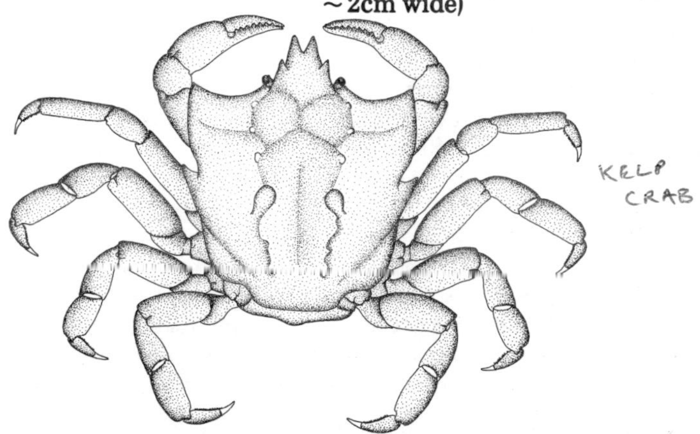

KELP CRAB

67. *Pugettia producta* (carapace ~ 3cm wide)

several centimeters beneath the sand, but they come to the surface to feed when waves wash over their habitat. They then partially bury the back end in the sand, leaving the head and front part of the body exposed and angled towards the sea. The large feathery antennae are unfurled in the backwash of the waves and filter a variety of food particles from the receding water. Because of these habits, *Emerita* is most easily collected at higher tides by digging into sand as the water flows seaward after a breaking wave has spent its energy. They burrow very rapidly when disturbed, so one must be quick with a shovel or hands. You can watch how they dig by putting one on sand in your hands or a bucket.

The regular or true crabs belong to the Section Brachyura and are among the most abundant and characteristic inhabitants of the intertidal region. They are found in their greatest numbers along rocky shorelines, although several species inhabit bay tidal flats. The vast majority of intertidal crabs are omnivores, primarily scavenging for their food. The larger forms may be active predators, and the unwary collector soon learns which species cannot be handled carelessly. The persistent visitor to local shores will undoubtedly turn up more brachyurans than the 10 species mentioned here, as we have included only the most common forms.

Two types of "pea crabs" (family Pinnotheridae) are locally abundant. *Pinnixa* and *Scleroplax granulata* are always found in association with other animals. The first is a symbiont within the mantle cavities of bivalves, especially *Tresus*, the horseneck clam (fig. 115). These pale whitish or pinkish crabs have rather weak appendages and depend in part on the clam's water currents to supply their food. The female crabs are larger than the males, with fat oval bodies. *Scleroplax* inhabits the burrows of the echiuran worm *Urechis* (fig. 44) and the ghost shrimp, *Callianassa* (fig. 60), in bay mud flats. Its body is laterally elongated allowing it to maneuver in the narrow tunnels of its hosts, moving sideways in typical crab fashion. This pea crab depends upon the water currents produced by the hosts to obtain food.

The remainder of the local brachyurans may be conveniently divided into three distinct groups, the spider crabs (family Majidae), the cancer crabs (family Cancridae), and the shore or grapsoid crabs (family Grapsidae). The first category includes two relatively common species. *Loxorhynchus crispatus* (fig. 66) is a low intertidal and subtidal decorator crab. Its habit of covering its body with living algae, hydroids, sponges, etc., earns it first prize as local camouflage artist. In addition, these crabs situate themselves in habitats which match their cryptic assemblage and, being quite sluggish, are very difficult to see among the growth in low rock pools. The exact function of this decorating by *Loxorhynchus* has yet to be firmly established, but predator avoidance is the most likely possibility. *Pugettia producta* (fig. 67) is a relative of *Loxorhynchus* but generally depends on protective coloration rather than decorating behavior for camouflage. Its habit of clinging to large brown algae gives *Pugettia* the common name of kelp crab. The body and legs are smooth, generally dark

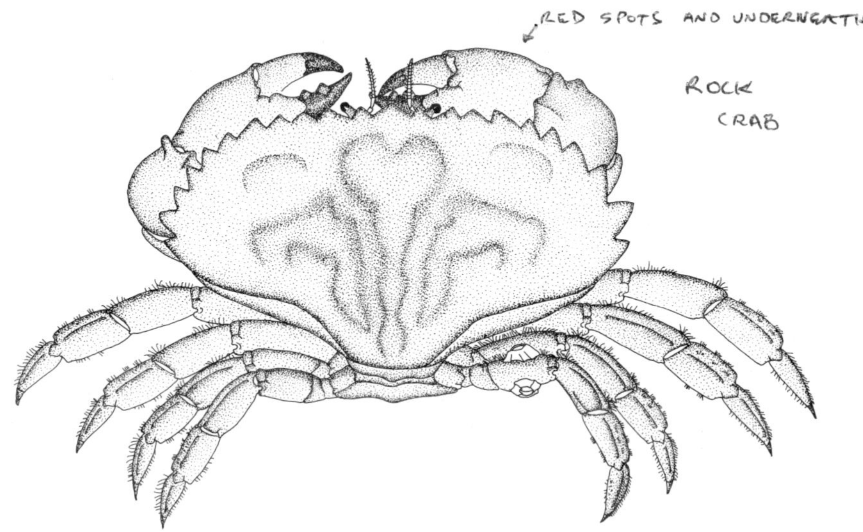

68. *Cancer antennarius* (~15cm wide)

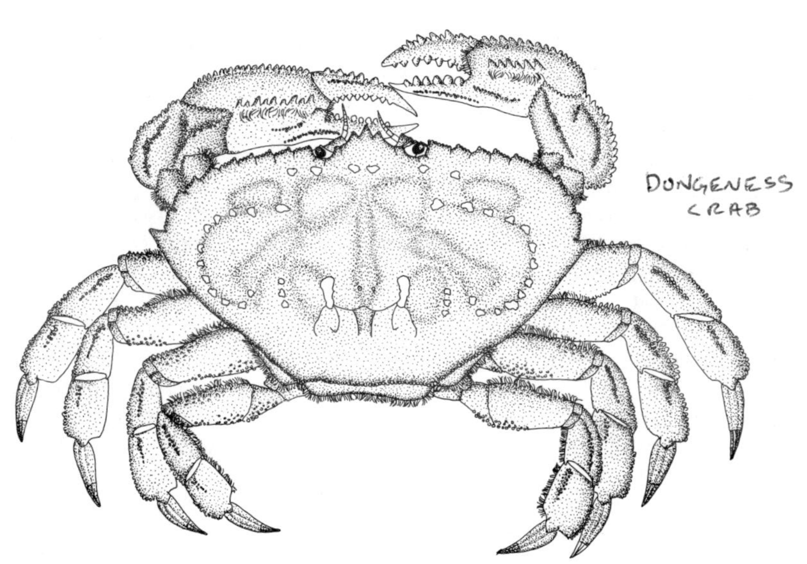

69. *Cancer magister* (~20cm wide)

reddish-bown or olive green in color (resembling the algae on which it is usually found). *Pugettia* is a much more active crab than *Loxorhynchus* and is extremely strong in spite of its thin legs and relatively small claws. Adults may measure several centimeters across the carapace, and the claws and sharped-tipped walking legs can cling tenaciously to kelp or a careless hand.

Three species of cancrid crabs may be encountered along the north coast. *Cancer antennarius* (fig. 68) and *C. productus* are both characterized by black-tipped claws, but the former is generally smaller and features red spots on the claws and underside of the body. Both of these species may be found in low tide pools or partially buried in sand under rocks. *Cancer antennarius* is the more common of these two species. The third species you are likely to find is the well-known and commercially important Dungeness crab, *Cancer magister* (fig. 69). While all three species are edible, only *C. magister* is large enough and occurs in high enough numbers in fishable waters to make it a profitable market species. In some areas of its range, *C. magister* has become quite scarce because of excessive exploitation and perhaps pollution. The bulk of the Dungeness crab population occurs offshore on sandy bottoms during most of the year. However, during the summer months the crabs may be found in Humboldt Bay where they come to molt in the protection of areas such as eelgrass beds. Many cast skeletons and "soft-shelled" individuals may be seen during this period.

The three common species of shore crabs (family Grapsidae) may be easily identified on the bases of habitat and coloration as described in the key. The lined shore crab (*Pachygrapsus crassipes*, fig. 70) is characteristic of the rocky mid- and upper-intertidal areas. This crab sports many fine green and red lines across the top of its 3-4cm wide carapace. Look for *Pachygrapsus* in cracks and crevices at about the level of mussel beds, or amongst the mussel assemblage. Lower in the intertidal, in pools and rock rubble areas, lives *Hemigrapsus nudus*, the purple shore crab (fig. 71). The carapace is deep purple and the claws are spotted red. *Hemigrapsus* and *Pachygrapsus* are both scavengers and tend to fill that function at their respective tidal levels, largely avoiding competition with one another. A third species of grapsoid crab is the smaller, greenish, bay shore crab, *Hemigrapsus oregonensis*. The high tolerance of this species to lowered salinities and turbid water allows these crabs to dominate in bay and brackish water situations. They may be found in large numbers under stones in the mud along the shores of Humboldt Bay and in mud burrows in the banks of various estuaries.

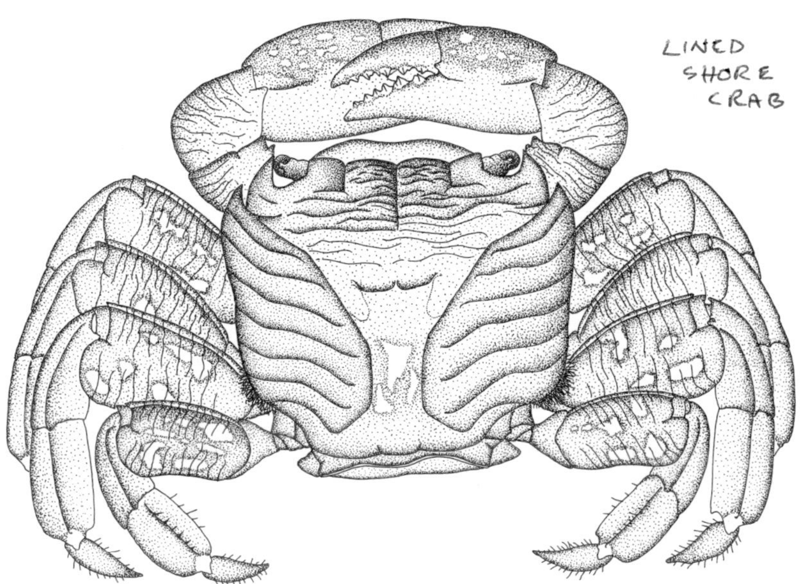

70. *Pachygrapsus crassipes* (carapace ~ 3cm wide)

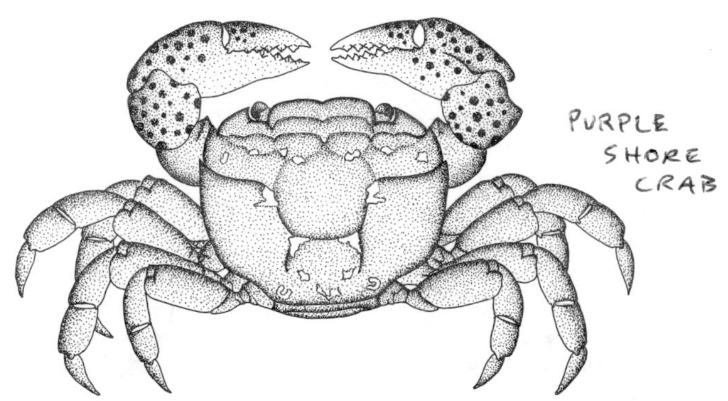

71. *Hemigrapsus nudus* (carapace ~ 3cm wide)

PHYLUM MOLLUSCA (soft-bodied)

the snails, clams, octopuses, etc.

The mollusks are probably the most familiar group of marine invertebrates in the world. Their diversity is exceeded only by that of the arthropods. This populous and widespread assemblage of animals is represented in nearly every environment that exists and is well known in the fossil record, having left traces of their presence from as long ago as 500 million years. These bilaterally symmetrical animals have, with few exceptions, taken up rather sedentary life-styles. The mollusks are non-segmented and have an open circulatory (hemocoelic) system similar to that of arthropods. The octopuses and squids are mollusks with active life-styles and have gone about the business of movement in new and unique ways. They are, thus, rather atypical as mollusks go.

As in most invertebrate groups one quickly notices one or two particular traits that stand out as being "feature items" of that group. With the mollusks the most obvious characteristic is the shell, and it is primarily upon the basis of this highly variable structure that the species of most molluscan types are differentiated. Modern taxonomists, however, are placing more and more emphasis on the soft innards, much to the dismay of those who maintain and curate large shell collections. In addition to the tendency to produce a pretty shell, mollusks are united in more subtle ways. They all have some sort of muscular "foot". This structure ranges from the obvious crawling foot of snails, to the digging foot of clams; exactly what has happened to the foot in squids and octopuses is not entirely clear, but we assume that it is there in some disguised form. Most mollusks also possess a fold of the body-wall tissue called the **mantle** which secretes the shell. The fleshy lining of a clam's shell is an example of this structure. This folding of the body wall results in a cavity or space between the mantle and the body mass itself called the **mantle cavity**. In typical cases this mantle cavity contains the gills, the anus, the excretory pores, and the reproductive pores. The evolutionary changes in the position, size, shape, and contents of this cavity which have taken place make a lovely set of stories of adaptation and design.

The embryogeny of most mollusks indicates some sort of evolutionary relationship between this group and the annelids. A large number of mollusks passes through a trochophore larval stage which closely resembles that of some polychaetes. Rather than elongating and segmenting, however, the molluscan trochophore develops a shell gland and is usally transformed into a compact **veliger larva**, unique to this phylum. The veliger has a tiny shell and a pair of large ciliated lobes for swimming. As is so often the case in a diverse group, the extreme variability of adult form is frequently reflected in developmental differences so that many patterns of larval life exist within the Mollusca.

Members of this phylum constitute a major portion of the total energy and living stuff (biomass) available to almost all marine communities, and thus they must be considered one of the most important phyla in the sea. Of the seven classes of Mollusca, four are commonly encountered along the shore. These are the Polyplacophora (chitons or sea cradles), Gastropoda (snails, limpets, and sea slugs), Bivalvia (clams, mussels, oysters, and scallops), and Cephalopoda (octopuses and squids). The members of the last class are generally subtidal, but at least one type of octopus may be seen intertidally if one looks long enough and carefully enough (or if one is lucky).

Class Polyplacophora
the chitons

The chitons are characterized by possessing a well-developed muscular foot designed for crawling and adhering to hard substrates (much like the foot of a snail), and a set of eight shells or **valves** arranged in a single line along the dorsal surface. The shells overlap like shingles, allowing most types to roll up, head to tail, in a posture reminiscent of pill bugs. In all but one species these shells are visible although sometimes overgrown with algae and bryozoans. When a chiton dies its shells are strewn about by the tides and waves; beachcombers recognize these as "butterfly shells". The fleshy part of the mantle forming a border around the margin of the back and at least partially covering the shells is termed the **girdle** and is of utmost importance in specific identification.

Chitons are, for the most part, grazers on algae and use a belt-like strap with file-like teeth called the **radula** for scraping plant material off rock surfaces. Many snails employ this same sort of structure for feeding. You may have seen terrestrial ones, slugs, or aquarium snails scraping algae off of glass. Polyplacophorans are inactive and graze within a relatively small feeding range. Many cling firmly to rocks and are difficult to remove without force. Do not try to force them from a rock, as you may injure them if they do not come off easily. The gills are located in a pair of grooves (the mantle cavity) one on each side of the body between the foot and the mantle edge. A water current moves from front to back along these grooves, over the gills, excretory and reproductive pores, and the posteriorly placed anus. The water then flows away from the animal out the back end of the grooves. These gill grooves are easily seen, and should be searched carefully (in large chitons) for a small commensal polychaete worm, *Arctonoe*, which feeds upon bits of material caught up in the water current. Gametes are also carried to the water along these grooves. Fertilization is external and a free-swimming larval stage develops which then settles as a young chiton.

Only a few chitons are identifiable in the field without microscopic examination of the nature of the girdle and valve surfaces. Even the simple key below will require the use of a hand lens to examine the surface texture of the girdle.

Key to Some Common North Coast Chitons

1. Large, usually over 20cm long; girdle completely covering shells; dorsal surface uniformly reddish-brown, warty. . . . . *Cryptochiton stelleri* (fig. 72)
1. Less than 15cm long; at least part of the shells visible along the dorsal surface. 2
2. Valves only narrowly exposed along the dorsal midlines; girdle black, slick, and leathery. . . . . *Katharina tunicata* (fig. 74)
2. Girdle relatively narrow, exposing most of the width of the valves; girdle not black, may be rough or smooth but not slick. 3
3. Small, usually less than 3cm long; girdle smooth; shells marked with pink and white, and often with thin blue lines. . . . . *Tonicella lineata* (fig. 75)
3. Girdle not entirely smooth; shells not marked as described above. 4
4. Front of girdle produced forward to form a large flap, often red beneath. . . . . *Placiphorella velata* (fig. 73)
4. Girdle not produced forward as a flap. 5
5. Girdle covered with large bristles or hair-like processes. . . . . *Mopalia* (fig. 76)
5. Girdle not covered with obvious bristles or hairs. 6
6. Girdle covered with tiny bumps, grainy in appearance. . . . . *Cyanoplax dentiens* (fig. 77)
6. Girdle covered with tiny overlapping scales (may require hand lens to see them). . . . . *Ischnochiton, Lepidozona* (fig. 78), and *Stenoplax* are genera which fit this description; further identification requires microscopic examination of shell surfaces.

Cryptochiton stelleri (fig. 72), the giant "gumboot" chiton, is a common inhabitant of semi-protected rocky shores from San Francisco north and, for its size, is a rather weak animal, often found loosely attached to rocks at low tides. The thick reddish-brown girdle completely covers the shells and gives this animal the general appearance of a wandering meatloaf. The butterfly shells from *Cryptochiton* are robust and survive the effects of erosion better than most; they are often found in sandy pools or on gravel beaches.

Placiphorella velata (fig. 73) is rarely seen along the north coast, and is restricted to fairly low intertidal zones, but is included here because it is so distinctive and unusual. Look for this animal on the undersides of rocks and ledges. This species is not a grazer on algae, but rather a predator, using its flap-like anterior mantle extension to clap down on unsuspecting prey. *Placiphorella* is thus an extremely aggressive animal as chitons go. The red color and frilled structure of the underside of the flap may attract prey such as small crustaceans or worms. The black "leather" chiton, *Katharina tunicata* (fig. 74) is most common on rocky shores subjected to rather heavy wave action. Look for this relatively large chiton within a narrow band in the upper-mid

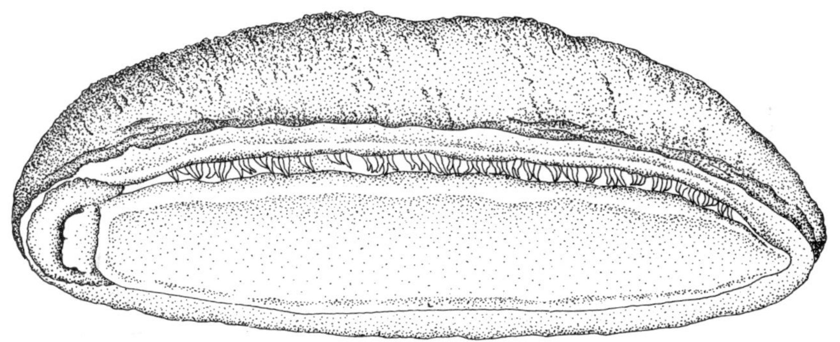

72. *Cryptochiton stelleri* (~20cm long)

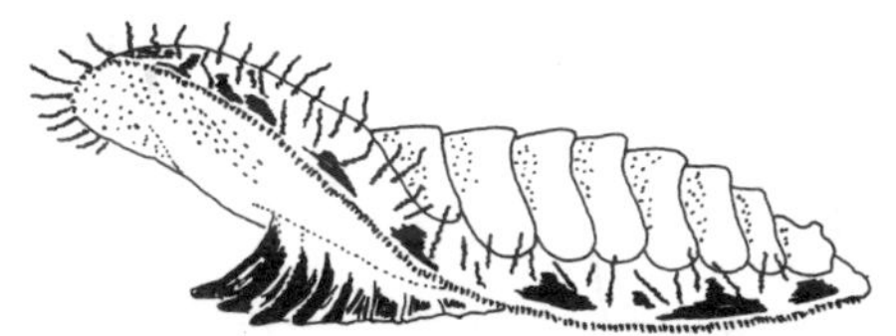

73. *Placiphorella velata* (~4cm long)

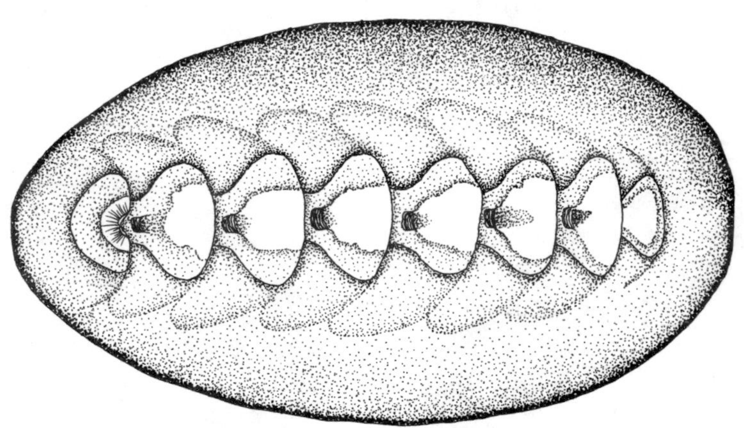

74. *Katharina tunicata* (~10cm long)

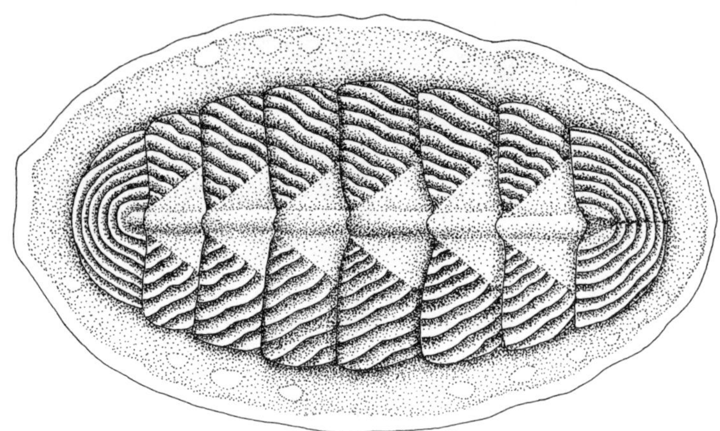

75. *Tonicella lineata* (~ 3cm long)

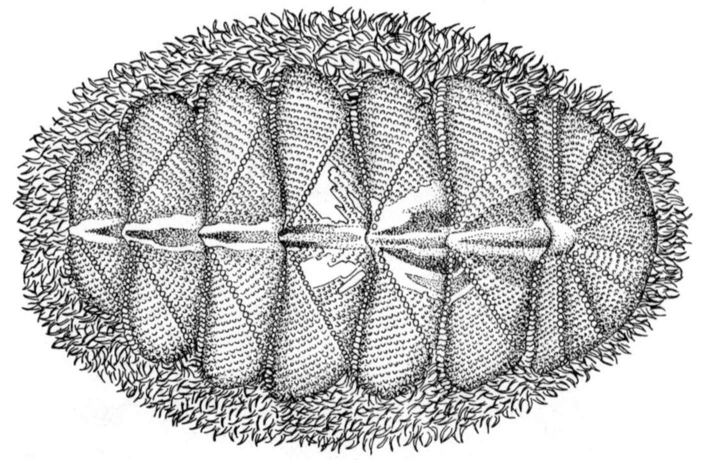

76. *Mopalia* (~ 6cm long)

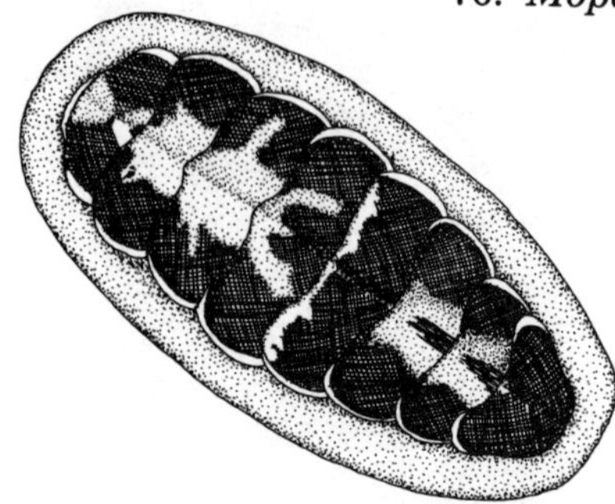

77. *Cyanoplax dentiens* (~ 4cm long)

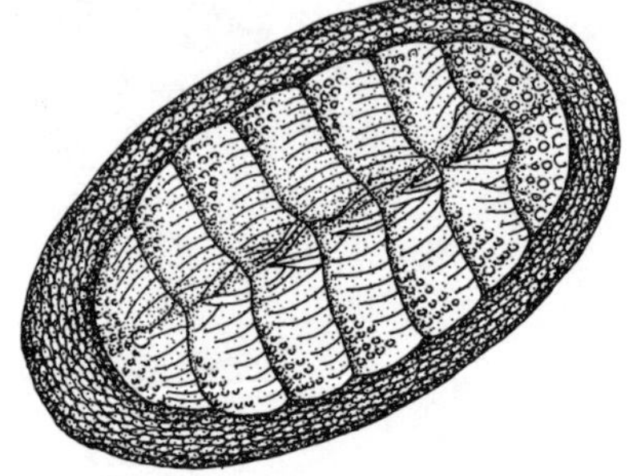

78. *Lepidozona mertensii* (~ 4cm long)

intertidal zone, usually on large rocks where the animals depend upon their tough bodies and firm grips for protection. *Katharina* is persistent in remaining at a particular level in the tide zone. When placed in an aquarium, members of this species will congregate on the sides of the tank at the water's surface. If the water level is lowered, *Katharina* will stubbornly keep its place and eventually dry up rather than crawl the few inches down to the water. In nature, of course, the return of the tides is more dependable than in the experimental tank of the zoologist, and the chiton's patience is invariably rewarded with periodic submersion.

Tonicella lineata (fig. 75) deserves special mention, for it is surely the most beautiful and one of the most distinctive of the Pacific coast chitons. The lovely pink and white, and sometimes blue, pattern on the shells makes the rest of the chitons quite drab in comparison. This mid-intertidal animal may be located in areas of moderate wave action living on the sides of boulders or occasionally under rocks. *Tonicella* often occurs on pink coralline algal encrustations, a habitat that somewhat matches the coloration of the shells.

Most of the other chitons are relatively secretive in their habits, usually locating themselves under medium-sized rocks in the mid- to low-intertidal zones. These include, among others, several species of the mossy chitons (*Mopalia*, fig. 76) and members of the genera *Lepidozona*, *Cyanoplax*, and *Ischnochiton* (figs. 77, 78).

Class Gastropoda

the snails, limpets, and sea slugs

The gastropods are the most diverse and successful group within the Mollusca. This class is characterized by the possession of a single shell, or no shell at all, a well-developed foot for crawling, and, at least in most shelled forms, having the mantle cavity with its associated structures located anteriorly over the top of the head. This last feature is a result of a larval process called **torsion** which involves a twisting of the body to bring the mantle from posterior to anterior. Thus the anus, gills, excretory openings, and reproductive pores are rather awkwardly positioned in front, causing a variety of sanitation problems for these gastropods. Keeping the gills from being fouled by waste materials, plus the unsavory habit of defecating over the head and feeding area, are problems which adult snails have had to face and solve through millions of years of evolutionary creativeness. The methods employed by gastropods to alleviate this situation tell an exiciting story of genetic plasticity and adaptation. Some forms such as the abalones and keyhole limpets have holes in their shells to allow a one-way flow of water through the mantle cavity; others have lost the gill from the right side of the mantle cavity and have displaced the anus and excretory pores to that area so that water may flow from left to right (over the gills first). The land snails and slugs have formed a "lung-like" air breathing chamber adapted to life on land.

The obvious question is, what were the reasons for torsion in the first place. What possible adaptive significance could this process have when it seems to present so many problems to adult gastropods? One likely answer is that torsion resulted in an advantage to the drifting, planktonic, larval stage which, of course, must survive if the animal is to reach adulthood. The late English zoologist, Walter Garstang (1868-1949) suggested that the placement of the mantle cavity over the head allowed the larva to retract its soft parts inside the developing shell to avoid being eaten by predatory plankters, although there are arguments against this hypothesis. Garstang expressed this idea in a verse about 1929 when he wrote "The Ballad of The Veliger, or How the Gastropod Got Its Twist", a poem published (after his death) in its entirety in a lovely little book called **Larval Forms and Other Zoological Verses** (Blackwell, 1966). The following two (non-adjacent) stanzas from "The Ballad of the Veliger" tell his hypothesis of the story. The first refers to the vulnerable, pre-torsional, planktonic veliger larva, and the second refers to the happy larva following torsion.

"But when by chance they brushed against their neighbors in the briny,
Coelenterates with stinging threads and arthropods so spiny,
By one weak spot betrayed, alas, they fell an easy prey—
Their soft preoral lobes in front could not be tucked away!

* * * * *

Predaceous foes, still drifting by in numbers unabated,
Were baffled now by tactics which their dining plans frustrated.
Their prey upon alarm collapsed, but promptly turned about,
With tender morsel safe within and the horny foot without!"

There are three subclasses of gastropods, two of which are well represented in the intertidal zone: the Prosobranchia include the typical shelled marine snails; the Opisthobranchia contain the often brightly colored sea slugs which generally lack a shell or have a small internal one; and the Pulmonata include the land snails and slugs. One small gastropod which occurs along our shore is placed within the Pulmonata by some workers; this is *Onchidella borealis*. It resembles a small limpet that has lost its shell, is brown or black, and rarely more than 1cm in length. This animal, like some pulmonates, has a hole opening into its mantle cavity near its tail end. It lacks gills in the mantle cavity and breaths air like pulmonates. Look for *Onchidella* amongst mussel beds or amongst algae on semi-protected rocky coasts.

Virtually all of the shelled marine gastropods are members of the subclass Prosobranchia. Of the many species you may encounter along our north coast, only a few are easily identified by means of obvious traits. Beyond these few species, the key identifies categories only, while we depend on the figures to help in further recognition of species in these categories.

Many other gastropods may follow through this key to the end, or apparently fall out at names whose illustrations and descriptions do not match the animal in hand. These problems are a result of the simplicity and incomplete nature of the key. We hope that readers who find the shelled gastropods especially intriguing will pursue identifications with one or more of the many fine shell books on the market (see references) or, better yet, with more professional published keys.

Key to Some Common Shelled Gastropods

1. Shell with one or more holes in addition to the main aperature from which the animal emerges. 2
1. Shell without additional holes. 4
2. Shell more or less flattened, with a row of several holes near the edge of the top surface (abalones). . . . . *Haliotis* (fig. 79)
2. Shell cap-shaped, not coiling, with a single hole at the apex. 3
3. Shell greyish with darker radiating lines; hole at apex more or less round. . . . . *Diodora aspera* (fig. 80)
3. Shell pinkish with darker brown or black radiating lines; hole at apex distinctly oval. . . . . *Fissurella volcano*
4. Shell cap-shaped or conical not showing any coiling (figs. 81-87). . . . . various limpets (see discussion)
4. Shell with at least some coiling. 5
5. Shell small, usually less than 1 cm long; shell dark brown or black. 6
5. Shell generally larger than 1 cm, colors various. 7
6. Shell generally black; found in high intertidal and spray zone. . . . . *Littorina* (figs. 90,91)
6. Shell dark brown; found in lower intertidal amongst algae. *Lacuna*
7. On bay tidal flats or sandy beaches. 8
7. On rocks or other solid substrates. 11
8. Shell large, up to 10cm in diameter, globose; shell chalky white with dirty brownish patches; on tidal flats in bays. . . . . *Polinices lewisii* (fig. 98)
8. Shell always much smaller than 10cm in diameter or length. 9
9. Shell very shiny and smooth, oval with small area of coiling evident at one end; grey to bluish in color; live snails on sandy substrates but shells inhabited by hermit crabs common in rocky intertidal. . . . . *Olivella* (fig. 97)
9. Not exactly as above in all respects. 10

10. Shell usually less than 2cm long with distinct black and white bands; on bay tidal flats. . . . . *Rictaxis punctocaelatus* (fig. 100) (This species is actually a shelled opisthobranch.)
10. Shell larger, up to several cm long, with low, close-set spiral and axial (along long axis of shell) ridges; a deep furrow around base behind shell aperture; up to several centimeters long. . . . . *Nassarius fossatus* (fig. 99)
11. Shell black or brown, more or less globular without a pronounced spire at the coiled end; shell relatively smooth without large bumps or ridges. 12
11. Shell not as described above. 13
12. Shell black; foot jet black on top and sides (sole is creamy white. . . . . *Tegula funebralis* (fig. 92)
12. Shell brown except where eroded to white; foot brown edged with orange. . . . . *Tegula brunnea*
13. With three prominent fan-like ridges along the axis of shell; relatively large, up to several centimeters. . . . . *Ceratostoma foliatum* (fig. 96)
13. Shell without prominent axial ridges. 14
14. Shell with a well-developed spike or tooth on the outer lip of the aperture. . . . . *Acanthina spirata*
14. Shell lacking a well-developed tooth on the outer lip of the aperture. 15
15. Aperture at least half as long as total shell length. . . . . *Nucella* (figs. 93,94)
15. Aperture less than half as long as total shell length. 16
16. Shell much longer than wide. . . . . *Searlesia dira* and others (fig. 95)
16. Shell nearly as wide as long; more or less top-shaped with numerous close-set spirals. . . . . *Calliostoma*

The following discussion includes only a few of the prosobranchs in the north coast intertidal zones. We have chosen as examples species which are large or common and which display some of the great variability in form and habit within this group.

Abalones (*Haliotis*, fig. 79) are rare in the intertidal areas of our coast because of pressure from sport and commercial fishing rather than from the habitat here being unsuitable. The delicate flavor of the large foot (when properly prepared), and the use of the shell for making trinkets and buttons, make these animals so highly prized that their numbers have been drastically reduced over the past few decades. In many parts of their range even the subtidal populations are seriously depleted. To those who would blame the sea otter for the lack of abalones we can only ask why the big snails didn't return when the otters were nearly rendered extinct. Both of these animals had certainly managed themselves well until humans became so efficient at killing them. At any rate, the seashore visitor will be lucky indeed to happen on an abalone these days. If you do find one, leave it alone and tell no one! The northern author has seen only a few in nearly a dozen years exploring the north coast, and is reluctant to reveal the exact locations of those sightings. There are several species of the genus *Haliotis* on the Pacific coast of North America. Three of these have ranges which include the northern California shoreline. *Haliotis cracherodii* (the black abalone) is more common to the south, *H. kamtschatkana* is frequently encountered well to the north, and *H. rufescens* (the red abalone) is the species most likely to be seen on local shores (fig. 79). The red abalone is primarily a subtidal form, but in rocky, wave-swept areas it may occur under ledges in the lower zones. The reddish shell of *H. rufescens* is rough, nearly corrugated, and often overgrown with algae and encrusting animals.

Abalones are herbivorous, scraping diatoms and other microscopic algae from the rocks with the large radula. These great snails crawl slowly forward, gently swinging their heads back and forth leaving an overlapping series of fan-shaped etchings where they have fed. Some species are also known to capture drifting blades of large seaweeds by reaching out the front of the foot, grabbing the alga and pulling it under the shell edge for subsequent dining. The large foot not only serves as the locomotory surface, but can clamp so tightly to the substrate that it is nearly impossible to dislodge the animal. The methodical use of pry bars and the availability of cold-water diving

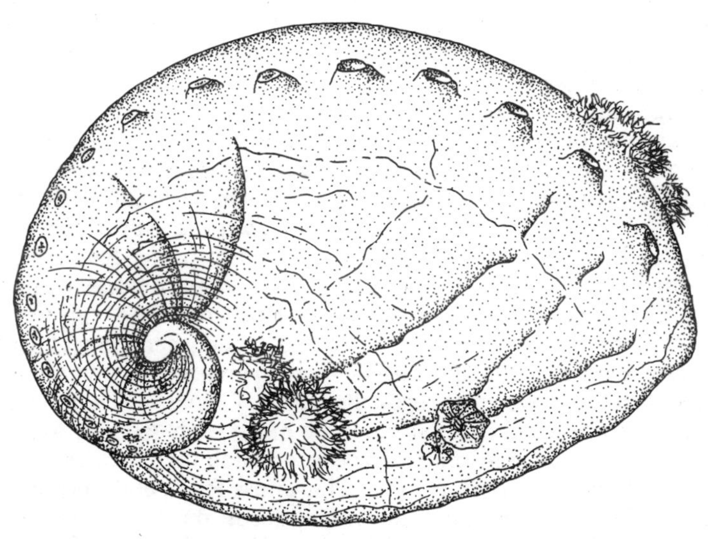

79. *Haliotis rufescens* (~ 18cm long)

suits have ended the days of intertidal abalone picking along the California coast by allowing shallow-water divers to reduce the near-shore populations thus reducing recruitment to the shore.

The **limpets** are among the most characteristic gastropods of the local rocky shores. Many of the various types display rather specific habitat preferences. Limpets, like the abalones, are basically alga-feeders, and their radula is adapted for scraping rock surfaces to that end. Identification of many limpets is not an easy task, and a wide range of color patterns and even shell shape may occur within a single species; only some have relatively distinct and non-variable features. Those chosen for inclusion here are usually recognizable by the nature of their shells or, in some cases, habitat restrictions. Not long ago one had a very high probability of correctly naming any limpet to the generic level by confidently calling it *Acmaea*. Sadly that name has gone the way of several others which were so familiar and comfortable to us as students. Now only one or two California limpets may be correctly addressed as *Acmaea*, the rest have been subjected to the effects of change in taxonomic status. Of the several limpets likely to be encountered along our rocky shores, two are distinctive in the possession of a hole at the point (apex) of the shell. These are *Fissurella volcano*, the volcano limpet, and *Diodora aspera*, the keyhole limpet (fig. 80). The hole in the shells provides an exit for water which has passed through the mantle cavity. Both of these species occur in the low intertidal areas of rocky shores, but *D. aspera* is by far the more common of the two. *Diodora* is an exception among limpets in its feeding habits since it is a carnivore feeding on such animals as bryozoans. Most limpets graze on algae. *Fissurella* is more common to the south but does occur nearly to the Oregon border. Look for this limpet on the sides and bottoms of low-intertidal boulders encrusted with coralline algae.

The high-intertidal to spray-zone areas of rocky shores are inhabited by two species of limpets which may compete for local food resources. The more common is *Collisella digitalis* (fig. 81) which often aggregates along cracks or depressions on the vertical faces of large rocks. Less frequently seen is *Collisella scabra* (fig. 82) which, at least locally, tends to be more solitary and seems to prefer less steep rock faces than *C. digitalis* (a behavior that may partially alleviate competition between the two species). Both of these species move down into the water at high tide to feed, and then retreat to their upper-zone levels as the tide recedes, and thus reduce the amount of time spent in the lower, more crowded zones. *Collisella scabra* is known to "home" to precisely the same spot after each feeding excursion. Each limpet forms a scar on the rock into which its shell fits perfectly, allowing a tight seal at low tide. There are four relatively common species of dark-colored limpets which occur on rocks through the upper-mid- to lower-mid-intertidal zones. *Collisella pelta* (fig. 83) is a fairly high-shelled limpet common among mussel beds and on rocks at about the level of the mussels, although it is sometimes found in

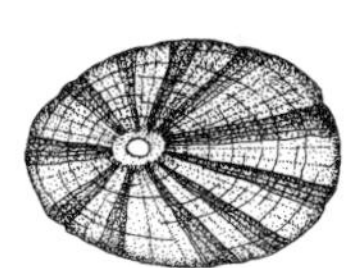

81. *Collisella digitalis*
(~1.5cm long) RIBBED

80. *Diodora aspera*
(~4cm long) KEYHOLE
CARNIVORE

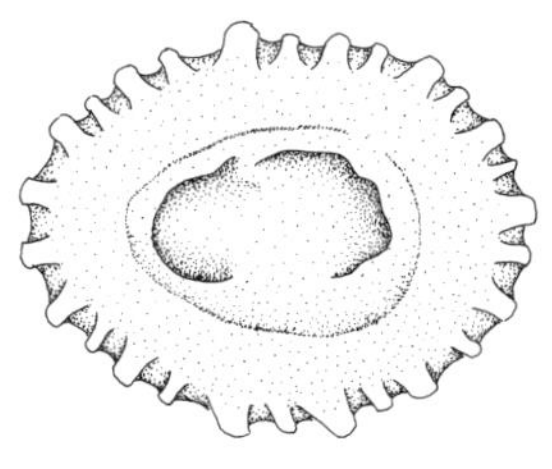

83. *Collisella pelta*
(~2.5cm long) SHIELD

82. *Collisella scabra*
(~1.5cm long) ROUGH

LOTTIA GIGANTEA
OWL

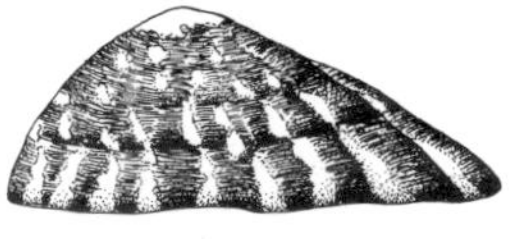

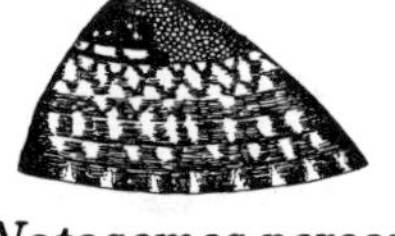

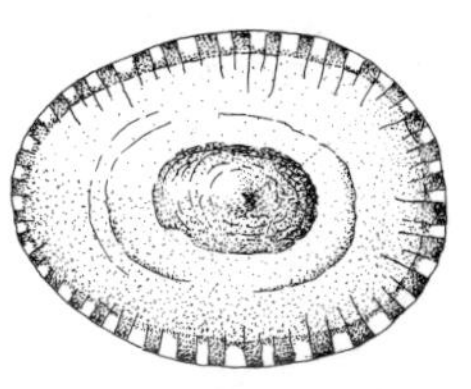

85. *Notoacmea persona*
(~3cm long)

86. *Notoacmea scutum*
(~2cm long) PLATE

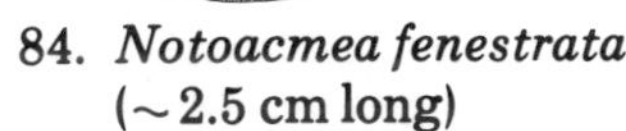

84. *Notoacmea fenestrata*
(~2.5 cm long)

CREPIDULA ADUNCA
HORNED SLIPPER

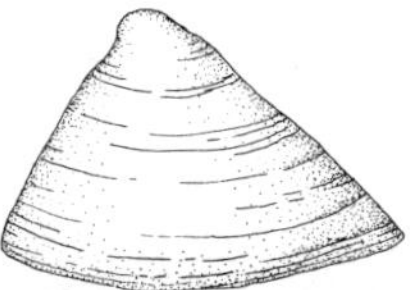

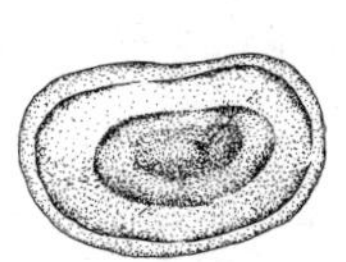

87. *Acmaea mitra*
(~2cm long)
DUNCE CAP

88. *Notoacmea incessa*
(~1cm long) KELP
FEATHER BOA

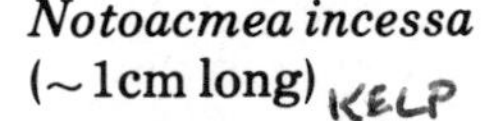

89. *Notoacmea paleacea*
(~5mm long)
PHYLLOSPADIX

higher zones. The shell often bears alternating light and dark radial lines, but the pattern is variable. *Notoacmea fenestrata* (fig. 84) is similar in shape to *C. pelta* but generally lacks the radiating color pattern. This species is frequently located on smooth boulders which are resting on sand in the mid- to low-intertidal zones. *Notoacmea fenestrata* often crawls to the lower sides of the rocks at low tides and actually buries itself beneath the sand. Dig the sand away from such stones and you may have a better chance of seeing this animal. A common species in the upper-mid zone is *Notoacmea persona* (fig. 85). The dark brownish shell is distinctly marked with a lighter checkerboard pattern, especially around the edges of the shell aperture. Look for *N. persona* in deep cracks and depressions relatively high on the rocks along semi-protected shores. *Notoacmea scutum* is another rocky shore limpet and may be recognized by its low profile (fig. 86). The aforementioned checkerboard pattern is also evident on this species, but it is often masked by a growth of green algal film over the surface.

Acmaea mitra (fig. 87) is commonly called the "dunce-cap" limpet. It is most common among boulder rubble in the low zones of wave-swept shores. The area south of Patrick's Point is a good place to search for this species. Its characteristic tall white shell is often covered with an encrusting pink coralline alga. Three local limpets are identifiable on the bases of their rather specific substrate preferences. *Notoacmea incessa* (fig. 88) lives and feeds upon the stipes of the feather-boa kelp *Egregia menziesii*, and the limpet's size and shape are well adapted to this habitat. One can often find *Egregia* with scars or hollowed-out depressions where the limpets have been feeding. A more spectacular adaptive form can be seen in *Notoacmea paleacea* (fig. 89). This small limpet lives exclusively on the blades of *Phyllospadix*, the surf grass, and its tall, very narrow shell fits perfectly along the narrow leaves. *Collisella instabilis* is more common south of here, but can be found locally with a little searching. This limpet resides on the rather cylindrical stipes of the brown kelp, *Laminaria*. The sides of the aperture are shaped to fit snugly downward like rockers over the edges of the kelp stipe. Because of this deformation of the shell, the animal will not rest firmly when placed on a solid, flat surface but rather will rock like a hobby horse at the slightest touch; hence, the specific epithet—*instabilis*.

The rocky intertidal spray zone is the usual habitat for two species of periwinkles. *Littorina planaxis* (fig. 90) is generally somewhat larger than *L. scutulata* (fig. 91), although neither type much exceeds 1cm in the height of the shell. While both are relatively common in rocky areas, *L. scutulata* can also be found on pilings and docks in bays. These small black or dark-brown snails are extremely tolerant of near-terrestrial conditions and, in fact, cannot withstand prolonged immersion in sea water. Yet, they can never stray inland from their spray-zone environment and thus are restricted to a very narrow strip of coastline. Littorines feed by scraping certain algae, particularly diatoms, from the rock surfaces. They are only active when the humidity is relatively high, for their gills must be kept moist. During dry conditions, they clamp down tightly against the substrate, close a protective door (**operculum**) across the aperture, and seal the shell opening to the rock with a glue-like mucus. Northern California littorines reproduce from about April or May until at least mid-October. Their elegant, microscopic egg cases are set adrift in the sea; the veliger larvae eventually break free from the cases to drift inshore and later settle as tiny snails.

Members of the genus *Lacuna* are small dark brown snails somewhat similar in size and appearance to *Littorina*, but occuring much lower in the intertidal. There are at least two local species of *Lacuna* (probably *L. marmorata* and *L. porrecta*), but the taxonomy of the genus has historically been the subject of some confusion and disagreement. Look for *Lacuna* on algae and surf grass in mid to low rocky intertidal pools, or on eelgrass in Humboldt Bay. These snails are extremely abundant throughout the spring and summer months during which time they lay small globular greenish, or doughnut-shaped white, egg cases on the plant surfaces.

No snail is more characteristic of the rocky California outer coast than the black turban *Tegula funebralis* (fig. 92). This alga-grazer may occur in extremely high numbers in the upper-mid zone pools of nearly any rocky area with moderate wave action. The abundance of these creatures is reflected in the high percentage of hermit crabs inhabiting their shells, probably more a matter of availability than of preference. Research on *T. funebralis* has shown that, in at least some parts of its range, the young snails settle from the plankton in the high intertidal region where they live for five or six years before moving lower down on the shore where food is more plentiful. Adults may reach an age of 30 or more years. *Tegula brunnea*, the brown turban snail, is far less common than *T. funebralis* and generally occurs in a lower zone, although the two species overlap a bit in vertical distribution. *Tegula brunnea* is typically larger than the black turban, often exceeding 3cm in shell diameter. Look for the brown turban along the Abalone Beach—Patrick's Point area or similar habitats, among boulders exposed by low tides.

Members of the genus *Calliostoma*, commonly called the top snails, are among the loveliest of the small (up to 2.5cm in diameter) snails. The shells of *Calliostoma* are usually brightly colored along the close-set spiral ridges.

There are two common species of "rock snails" on the northern California coast. *Nucella emarginata* (fig. 93) generally occurs throughout the mussel-bed zone. *Nucella lamellosa* is a larger animal, typically found in the lower- to mid-intertidal areas. Figure 94 illustrates two local forms of this species. As opposed to most of the other snails discussed thus far, these species are predators, feeding mainly on small mussels and barnacles by using the radula as a drill. *Nucella emarginata* lays its eggs in vase-shaped, yellowish cases which the animal attaches to rocks. No free-swimming larvae are produced,

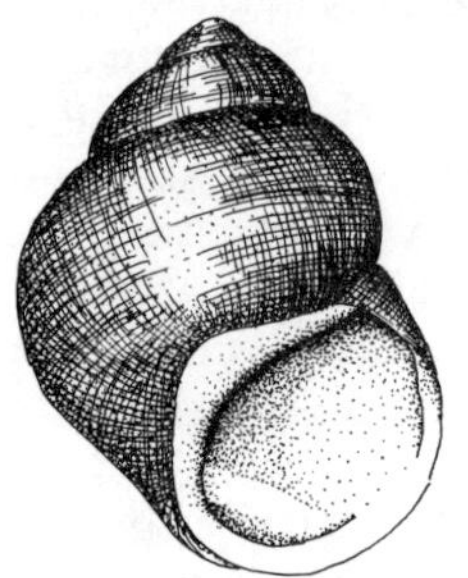

90. *Littorina planaxis*
(~ 1cm long)

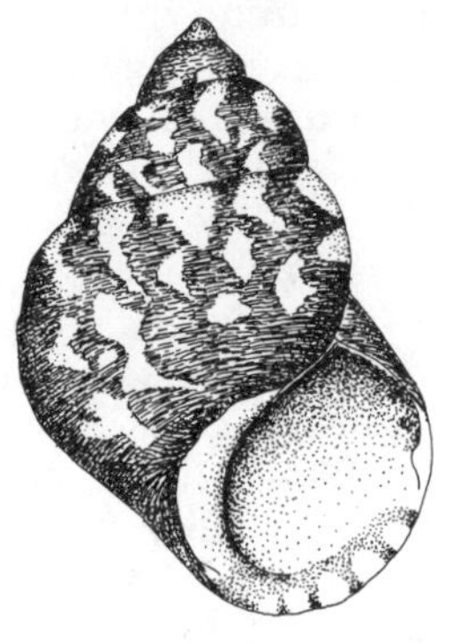

91. *Littorina scutulata*
(~ 1.5cm long)

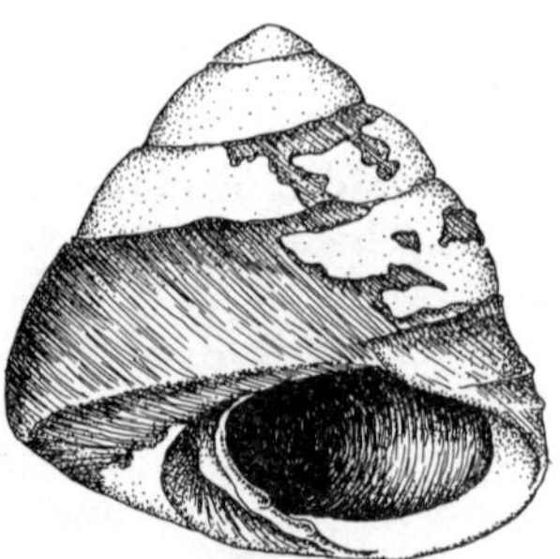

92. *Tegula funebralis*
(~ 2cm diameter)

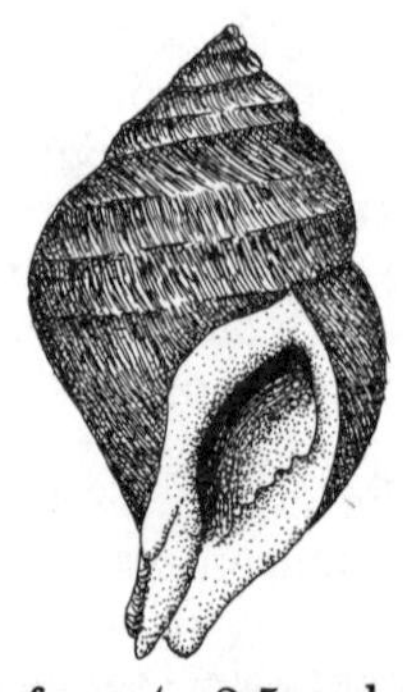

smooth form (~ 2.5cm long)
found in areas of heavy wave action

94. *Nucella lamellosa*

rough form (~ 4cm long)
found in protected areas

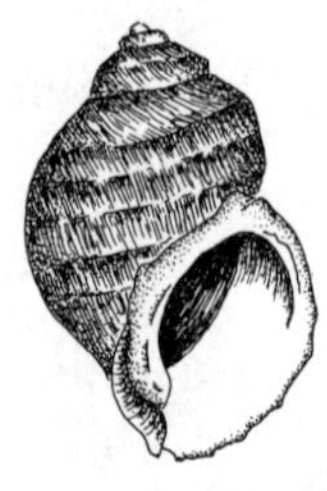

93. *Nucella emarginata*
(~ 2cm long)

but rather the embryos develop directly into tiny snails within the egg case. Although several eggs are depositied within each case, only a single juvenile emerges, that one having devoured its siblings during development. Incidentally, until recently these snails were known as *Thais*, another familiar name now fallen into disuse. *Acanthina spirata* superficially resembles *Nucella* and also occurs in and around mussel beds. *Acanthina* is distinguishable by the presence of a pronounced spike on the shell aperture. It was once thought that the tooth functioned in chipping apart or prying the shells of small barnacles upon which the animal feeds. As it turns out, however, *Acanthina* apparently drills its prey with the radula, and the function of the tooth is not clearly understood. *Acanthina* is not as common as *Nucella* along our coast. At least two other species of relatively large prosobranch gastropods may be commonly encountered along our rocky intertidal areas. *Searlesia dira* (fig. 95) is a 4-5cm long snail which lives in cracks and depressions on rocks and gravel areas of the mid- to low-intertidal zones. The shells of *Searlesia* often house hermit crabs living in the upper tidal zones. *Ceratostoma foliatum* (fig. 96) is one of the largest local gastropods, reaching 7cm or more in length and bears perhaps the most ornamental shell of all the local snails. The flap-like ridges may serve to increase the strength of the shell or blend in with the irregular surface of rocks on which the animal lives. Search for *Ceratostoma* among boulder rubble and under rocks which are not imbedded in sand.

The olive shells (*Olivella*) typically occur on low intertidal to subtidal, protected, sandy bottoms. The expanded foot and elongate water siphon allow the animal to crawl through the surface of soft, unstable substrates partially buried. There are two species of local olive shells, *Olivella pycna* (fig. 97) and *O. biplicata.* The latter is more common subtidally and to the south; the former may be encountered on low intertidal sand flats in Humboldt Bay. *Olivella* generally occur in very large populations and the empty shells are commonly carried up and down the coast by local currents and wave action. Consequently, you may often find their shells in places other than their normal habitats such as rocky shores where they are frequently inhabited by hermit crabs.

Except for the abalones you will find no larger local snails than *Polinices lewisii* (fig. 98), the moon snail. The large, nearly spherical, chalky white or dirty brownish shells may reach the size of baseballs. *Polinices* is locally restricted to bay mud and sand flats, and is relatively common. The huge foot allows the animal to crawl about over soft substrates, more or less plowing through the surface sediments. When alarmed or irritated, the water-filled foot is quickly drained and fully retracted into the shell, pulling the operculum in afterwards like a manhole cover to close the shell opening. Unfortunately we have noticed a marked decrease in the numbers of Humboldt Bay moon snails over the past year or so. While the shells are attractive enough and large enough to serve as containers for small house plants, empty ones are not uncommon and they are certainly plentiful and cheap in curio shops; therefore, we doubt that they are being killed for their shells alone. Rumor has it that some local residents have decided that certain parts of the moon snails are good to eat and that they are rapidly devouring the nearshore population. We hope that this fad will soon pass and that the natural-food enthusiasts will turn to more plentiful and renewable sources of nutrition, for a species so restricted in its intertidal habitat cannot long tolerate much harvesting pressure. Moon snails are carnivorous, drilling clean round holes in other mollusks or, more frequently, feeding on the carcasses of dead animals. The egg case of *Polinices* appears as a broken collar of sand up to 20 or more centimeters across. The eggs are released in a sheet of gelatin-like material covered with a sticky mucus to which sand grains adhere. The sand collar eventually breaks apart and thousands of larvae are released. As is so often the case in larval life histories, only a very few will survive to adulthood.

Nassarius fossatus (fig. 99), the basket shell, is characterized by a deep furrow around the base of the shell, and close-set spiral and axial ridges which give the shell the appearance of being covered with small bumps or nodes like the wickerwork of a basket. This predator and sometimes scavenger may reach several centimeters in length. *Nassarius* is common on bay tidal flats and is often in areas of eelgrass beds. This species lays its small (less than 1cm long) plate-like egg cases on the blades of the eelgrass, gluing them on by one edge so they extend vertically in the water.

Members of the Subclass Opisthobranchia include the often brightly colored sea slugs and their close relatives, many of which are certainly among the most beautiful animals of the sea shore. They occur in a wide variety of colors and patterns equalled only by the tropical fishes. They are particularly abundant and diverse along the north coast, and the careful observer should find several species during a morning's visit to the low intertidal zones. Nearly all opisthobranchs lack a shell or have the shell reduced in size and situated internally. One local species (*Rictaxis punctocaelatus,* formerly *Acteon*) does have a shell and is diagnosed as an opisthobranch by the traits of its soft parts (fig. 100). It is included in the above key to shelled gastropods. The beautiful black and white shells of *Rictaxis* are usually a centimeter or so long. Until 1971 this species was a regular summer inhabitant of Humboldt Bay, especially in the area of Elk River Slough, and the lovely shells were there in great abundance. Apparently they did not go unnoticed by others for we have seen none in recent years. They are generally recorded as occurring sporadically along the California coast, but we suspect that such a long-term absence is caused by something other than natural fluctuations. The following key and associated drawings will serve to help you identify the commonly encountered sea slugs of our area.

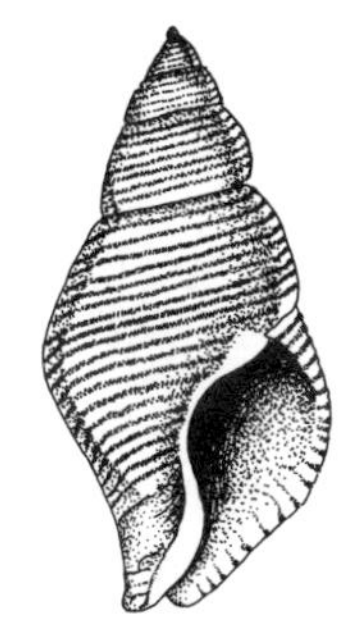

95. *Searlesia dira* (~3cm long)

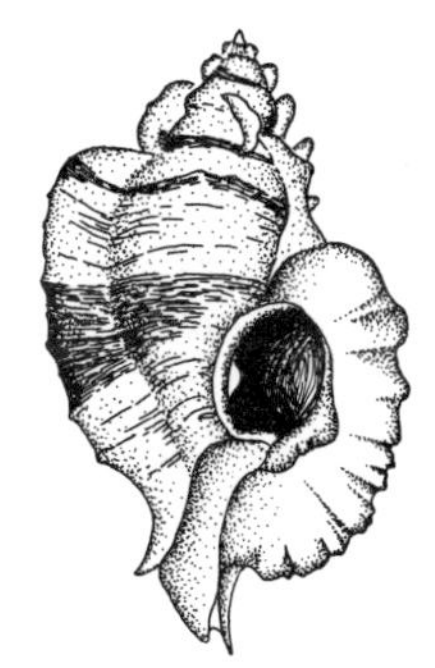

96. *Ceratostoma foliatum* (~5cm long)

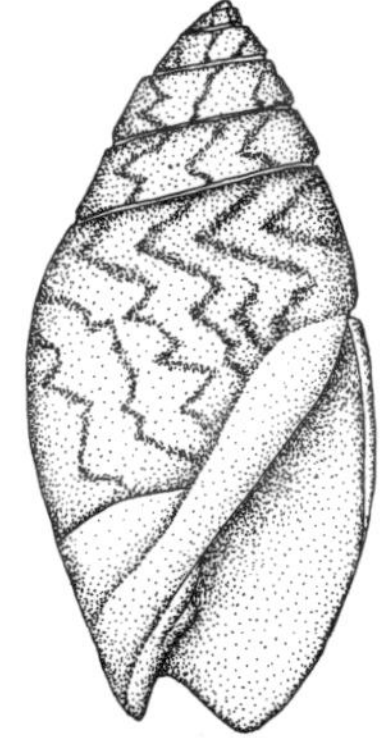

97. *Olivella pycna* (~2cm long)

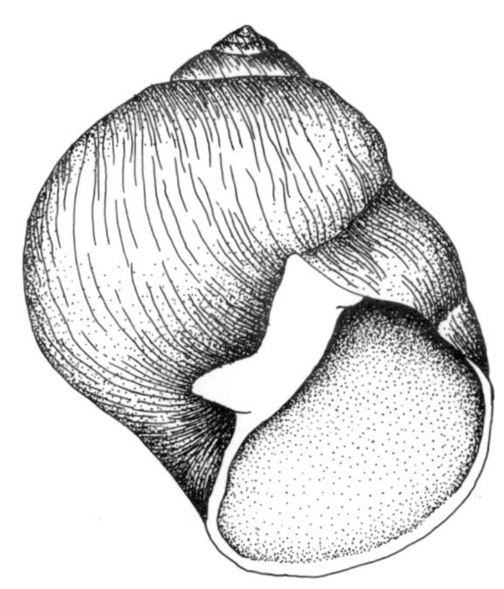

98. *Polinices lewisii* (~10cm diameter)

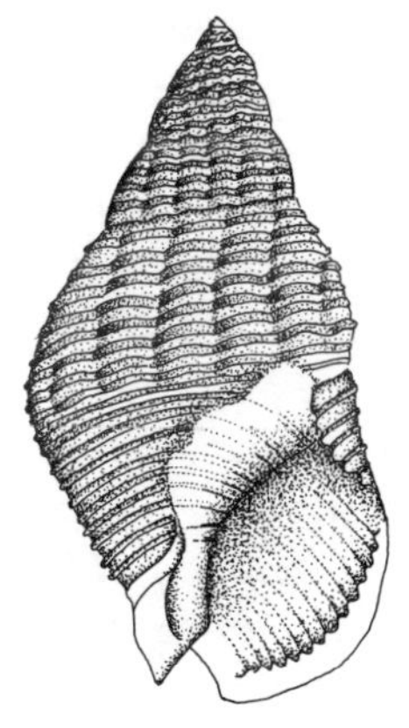

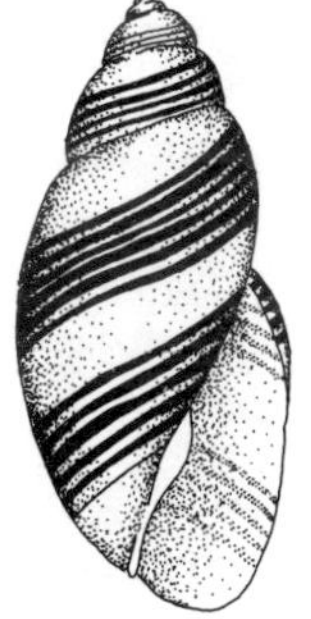

100. *Rictaxis punctocaelatus* (~1cm long)

99. *Nassarius fossatus* (~3cm long)

Key to the Common Opisthobranchs of Northern California

1. Body green with black lines; on eelgrass in bays .... *Phyllaplysia taylori* (fig. 101)
1. Body coloration otherwise 2
2. Body basically white, with bright orange bumps and short processes; no other obvious colors.... *Triopha carpenteri* (fig. 102) (The northern author reports that *T. maculata* may also (rarely) be encountered locally. This species is characterized by a dark brown body with small blue spots. One recently observed specimen housed the scale worm *Arctonoe* under the veil-like extension from the front of the head.)
2. Color pattern not as above 3
3. Back with numerous relatively long mobile processes (**cerata**) ... 4
3. Back smooth or with small bumps or warts, no cerata; a circlet of gills located posteriorly on the dorsum 10
4. Cerata distinctly branched (small animals, usually less than 2cm long).. 5
4. Cerata not branched (usually larger than 2cm, or if less than 1cm, bright red in color) 6
5. Color whitish, translucent, with distinct network of white lines on dorsal surface .... *Tritonia festiva*
5. Color and markings otherwise, usually dull brownish; small, common on hydroids........................ *Dendronotus* (fig. 103)
6. Body grey-white without bright markings; cerata dense giving animal a "hairy" apearance .... *Aeolidia papillosa* (fig. 104)
6. Body either with bright markings or, if white, cerata arranged in distinct patches or tracks and not extremely dense over dorsal surface.. 7
7. Cerata relatively thin, finger-like 8
7. Cerata flattened, relatively broad and leaf-like or blade-like...... 9
8. Body with orange and white markings on a translucent background, with bright, "neon" blue, thin lines on surface between two main rows of cerata ...... *Hermissenda crassicornis* (fig. 105)
8. Roughly as above but without blue lines... *Antiopella barbarensis*
9. Body translucent white .... *Dirona albolineata* (fig. 106)
9. Body brownish.................................. *Dirona picta*
10. Small, usually less than 1cm long; uniformly bright reddish orange.................................... *Rostanga pulchra*
10. Usually much larger than 1cm in length; color never uniformly

reddish orange. 11

11. Body bright yellow with scattered brown or black spots 12

11. Color otherwise; surface smooth or with warts 13

12. Dark spots occur only between bumps on back . . . . *Anisodoris nobilis* (fig. 107)

12. Dark spots occur on and between bumps on back . . . . *Archidoris montereyensis*

13. Back relatively smooth, no obvious bumps or warts; color tan to light brown with darker chocolate brown solid or doughnut-shaped spots. . . . . *Diaulula sandiegensis* (fig. 108)

13. Back not very smooth; markings not as above 14

14. General body color tan or brown; anterior processes (paired **rhinophores**) and gill circlet tipped with maroon; warts yellowish. . . . . *Acanthodoris nanaimoensis*

14. General body color white; rhinophores and gills tipped with yellow; warts yellow; with a yellow line around edge of body. . . . . *Cadlina luteomarginata* (fig. 109)

With the exceptions of *Rictaxis* (classified as a cephalaspidean) and *Phyllaplysia taylori,* discussed below (which has an internal shell and is classified as an anaspidean), the common local opisthobranchs are within the Order Nudibranchia (having naked or exposed dorsal "gills"). These animals have lost the "true" ancestral gastropod gills and secondarily developed new gas exchange structures either as a plume-like cluster on the rear of the dorsal surface or as numerous finger-shaped or branched processes on the back called **cerata** (singular **ceras**). Nearly all opisthobranchs are carnivores, using the radula to scrape at sponge, ascidian, or bryozoan colonies, or to nip off indivi-

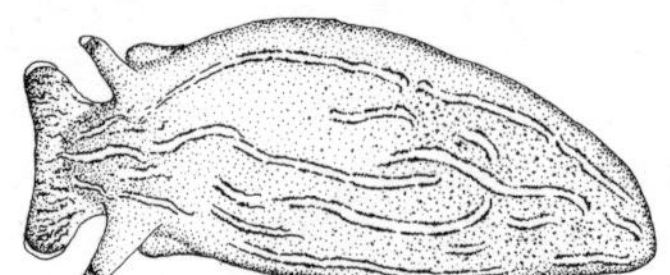

101. Phyllaplysia taylori
(~4cm long)

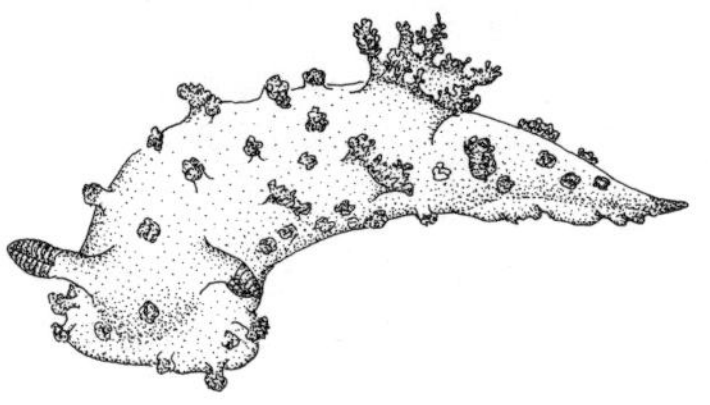

102. *Triopha carpenteri* (~4cm long)

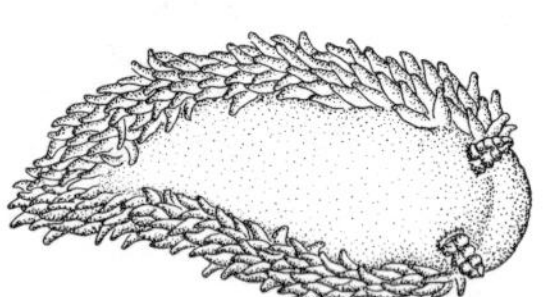

104. *Aeolidia papillosa*
(~3cm long)

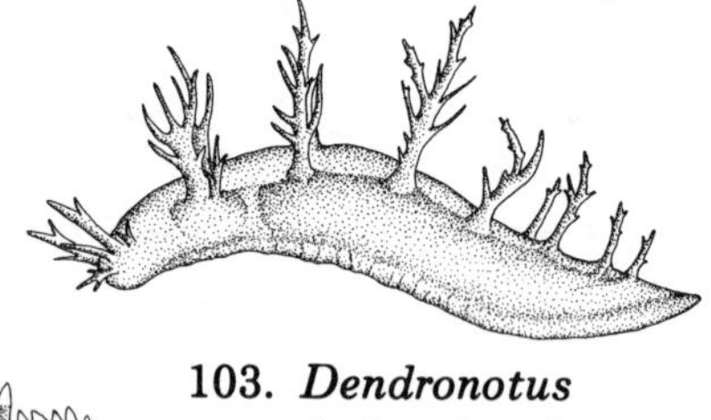

103. *Dendronotus*
(~1cm long)

105. *Hermissenda crassicornis* (~2.5cm long)

107. *Anisodoris nobilis*
(~7cm long)

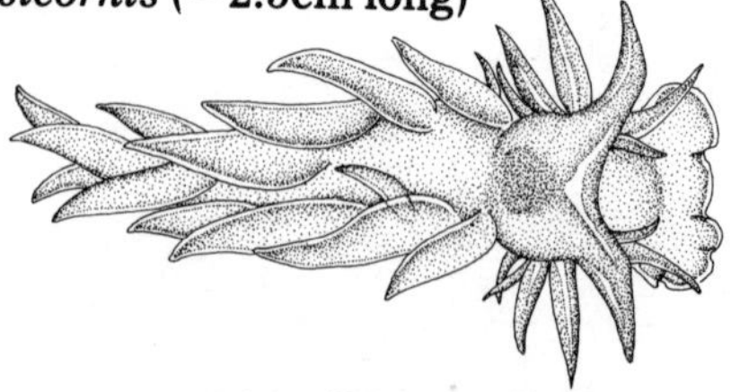

106. *Dirona albolineata*
(~3cm long)

108. *Diaulula sandiegensis*
(~5cm long)

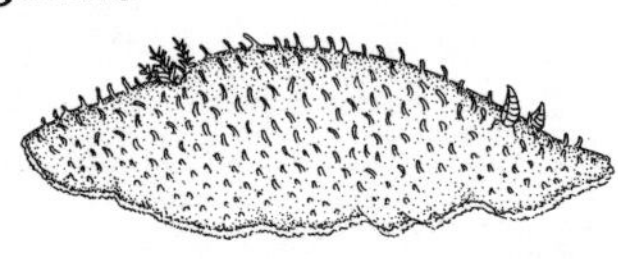

109. *Cadlina luteomarginata*
(~2.5cm long)

duals from clumps of hydroids. A few feed upon sea anemones. Some of the nudibranchs which eat cnidarians do not actually digest the stinging threads (nematocysts) of their prey. Remarkably these nematocysts do not become discharged and are moved to the tips of the cerata where they may function in defense of the sea slug. Exactly how this amazing feat is accomplished is unclear.

Nudibranchs of various sorts may be found in nearly all low-intertidal protected habitats. Look for them in deep cracks and crevices, under ledges, on wharf pilings, in tidal pools, on plant surfaces, and on potential prey species. In a few cases the coloration serves as camouflage. The small red nudibranch *Rostanga pulchra* matches precisely the color of the sponge upon which it normally lives and feeds; various species of *Dendronotus* (fig. 103) are small and match the color and branching form of the hydroid colonies upon which they are commonly found. *Phyllaplysia* (fig. 101), the aforementioned anaspidean, aligns itself on the blades of eelgrass in bays, its green body with black lines closely resembling the color and venation of the plant. More often, however, nudibranchs stand out like bright neon signs against the drab background of the rocks and seaweeds. That these animals appear to be advertising their presence is a phenomenon not completely understood. Some speculations exist regarding nematocysts and noxious skin secretions including acids. It has been observed that most fishes will eat nudibranchs but will immediately spit them out with great vigor as if the slug hurts their mouth or has a bad taste.

Class Bivalvia

the clams, including oysters, cockles, and mussels

The bivalves are so named because they have two shells (**valves**) enclosing the body. The shells are lined with the fleshy tissue of the mantle which forms the outer wall of the mantle cavity in which the gills are suspended. Bivalves are filter-feeders, drawing water into the mantle cavity through an **incurrent opening** or **siphon**, and forcing it out through an **excurrent siphon**. Within the mantle cavity the water passes over and through the highly specialized gills where food is trapped in sheets of mucus and transported to the mouth. Portions of the gills also serve a gas-exchange function. After passing through the gills, the excurrent water flow passes by the excretory pores and anus, carrying waste products with it as it goes out of the mantle cavity. The siphons are located posteriorly and may be elongate and fused together *(Tresus)* or separate *(Macoma)*, or no more than simple openings between the edges of the mantle *(Mytilus)*. The excurrent siphon is always of a smaller diameter than the incurrent one, so that water with its wastes leaves at a higher velocity and is propelled away from the body. The foot of bivalves is directed anteriorly and ventrally from the body. In some animals the foot is well developed and muscular for digging in soft substrates (clams), while in others it is highly reduced and rarely emerges from between the shells (mussels). Most bivalves release their gametes (sperm and eggs) into the water where fertilization occurs. In a few, however, the female takes sperm in with the incurrent water and internal fertilization occurs. The embryos are then brooded in special parts of the gills. Most species produce swimming larvae, looking much like gastropod veligers but with two shells, which eventually settle as juveniles.

In addition to the bivalves identified and discussed below, there are a few which are not in the key but which deserve separate mention. Several species of the Family Pholadidae may be encountered as burrowers in sandstone, shale, or even very hard calcareous substrates. The anterior end of the shells of boring pholads is often toothed or serrated, and aids in excavation by rasping out a burrow in the rocks. Once the burrow is formed, the siphons are extended out through the entrance into the surrounding water. Another type of boring clam is the so-called "shipworm." The common local species is *Bankia setacea*. You may find these animals in rotted pilings and planks which have long been submerged in sea water. *Bankia* looks like a worm, but close inspection reveals a pair of tiny shells at the back end. These animals, along with certain other wood-boring invertebrates (especially the isopod, *Limnoria*), cause tremendous damage to man-made wooden objects exposed to the sea. Beyond these odd types, most bivalves may be found either attached to hard substrates or buried in mud or sand.

Key to the Common Bivalves of Northern California

1. Attached to hard substrates (either cemented by one valve or attached by tough threads called **byssal threads**)................ 2
1. Not immovably attached to hard substrates, living in sand or mud 5
2. Cemented to substrate by one valve; shells with irregular rough sculpturing.. 3
2. Attached by byssal threads; shell more or less smooth and regular in outline and sculpturing, brown or black (mussels)............ 4
3. Shell more or less round in top view (rock scallop)............... .... *Hinnites giganteus* (fig. 113)
3. Shells elongate to irregular; grey to blue-black markings on whitish background (oyster).......... *Crassostrea gigas* (fig. 112)
4. Shell rough with well-defined growth lines; found on rocky shores (California mussel)............... *Mytilus californianus* (fig. 110)
4. Shell smoother; found in bays (bay mussel) *Mytilus edulis* (fig. 111)
5. Siphons fused into one long structure covered by a leathery sheath ("neck"); siphons so large that they are not fully retractible inside shells.. 6
5. Siphons variable, fused or separate, but can be fully retracted within shells.. 7

6. Posterior end of shells "cut" blunt where siphons emerge; large shells, up to 20cm or more in length (gaper clam) .... *Tresus* spp. (fig. 115)
6. Posterior end of shells evenly rounded; shells generally less than 15cm long (soft-shelled clam)............. *Mya arenaria* (fig. 116)
7. Shells elongate, much longer than wide; found on wave-swept, sandy beaches (razor clam)............... *Siliqua patula* (fig. 117)
7. Not as above in form or habitat.................................. 8
8. Shells with very prominent ridges radiating from area of hinge; shells broadly arched giving animal a rounded, globose appearance in end view; siphons short protruding less than 1cm from between shells; found near surface on tidal flats (heart cockle)....... .... *Clinocardium nuttallii* (fig. 118)
8. Radial ridges not especially pronounced in comparison with concentric lines, or radial ridges lacking altogether 9
9. Shells whitish and smooth, often with black markings; siphons long, thin, and separate; animal rather flat in end view .... *Macoma* (fig. 119)
9. Shell not very smooth nor flat in end view.................... 10
10. Large clams, to 15cm in shell length; shells with heavy concentric ridges but no radial ridges (Washington clam).................. .... *Saxidomus nuttalli* (fig. 120)
10. Smaller, to about 8cm in length; shells with concentric lines and weak radial ridges (rock cockle) .... *Protothaca staminea* (fig. 114)

There are two commonly encountered species of mussels along the California coast, *Mytilus californianus,* the abundant rocky-shore California mussel, and *M. edulis,* the bay mussel (figs. 110, 111). Both of these animals attach to hard substrates by byssal threads. The foot is much reduced and normally held inside the shells. At the base of the foot is a deep, pit-like byssal gland responsible for production of the tough, slightly elastic threads. The foot is used to place the threads on the substrate for attachment. The substance which forms the threads is at first a liquid which flows down a groove in the foot to the substrate. The material then hardens in the groove (which acts like a mold) to form this tough thread. The foot is then retracted and may be extended in another direction to attach a new thread. The pointed end of the shells is the anterior aspect of the animal and is generally directed downwards leaving the posterior siphonal openings directed up and slightly away from the substrate. The shells gape slightly when the tide is up, allowing water to enter and leave the mantle cavity in which food and oxygen are extracted by the gills. Areas of relatively heavy wave action are required by *Mytilus californianus* and apparently avoided by *M. edulis.*

Mytilus californianus is one of the most common inhabitants of California's rocky coastline. Large numbers of these animals may be seen on nearly any wave-swept area with a suitable substrate for attachment, and they form the "heart" of the characteristic "mussel-bed zone" or assemblage. They are commonly associated with the stalked barnacle (*Pollicipes polymerus,* fig. 49) and certain species of acorn barnacles (*Balanus,* fig. 48), but the dense mats formed by these animals provide housing for an incredible number of less obvious creatures. Worms, small sea cucumbers, limpets and other gastropods, small crustaceans, and several species of algae abound beneath, on, and between the mussels. Here they are provided with protected places to live and an accumulation of food material. The diversity of animal life associated with these mussel beds is largely influenced by the various sediments that are trapped between the mussels; the greater the number of sediment types, the greater the number of animal types. One of the main characters in the story of the California mussel is the common ochre seastar, *Pisaster ochraceus* (fig. 128). This animal is an active predator on mussels, and during low tides you may see these seastars on the rocks just below the mussel-bed zone. As the tide comes up, the seastars move into the areas of the mussels to feed. Look for paths of bare rock in the mussel beds, an indication of *Pisaster's* activities. In fact, as we have seen, the feeding activities of *Pisaster* help keep overall species richness high by preventing overdominance of beach habitats by *Mytilus.* In areas where *Pisaster* are absent (or removed by experimental manipulation), *M. californianus* eventually covers all available rock surfaces and extends its vertical distribution to depths as great as 10 meters where it may grow to enormous sizes.

Significant destruction of mussel beds along our coast has occurred at the hands of people. The large mussels are eaten and provide a tasty diversion from more common sea foods (at least when they are not carriers of shellfish poisoning). They are also used as fish bait as are certain large polychaetes which inhabit the mussel beds. While we do not argue against such activities as eating or fishing, the removal of animals from mussel beds can be accomplished in a variety of ways, some of which are highly damaging. A common practice is simply to rip off everything down to bare rock in huge chunks and then pick through the resultant heap for the desired creatures, leaving the rest to die on the beach. The northern author once stood next to two men engaged in such an activity while delivering a lecture to his class concerning the delicate nature of mussel beds. The only immediate result of this presentation was to direct the men's aggressive behavior away from the mussels and towards the author. Some experts estimate that an area denuded of mussels may take five or more years to return to "normal." With a little care, individual mussels may be taken from beds without drastically disturbing the rest of the assemblage (by prying them out with a suitable tool). Choose and take large mussels, one here and one there, selectively thinning the beds and giving smaller ones more room in which to grow. (Remember you need a

110. *Mytilus californianus* (~ 8cm long)

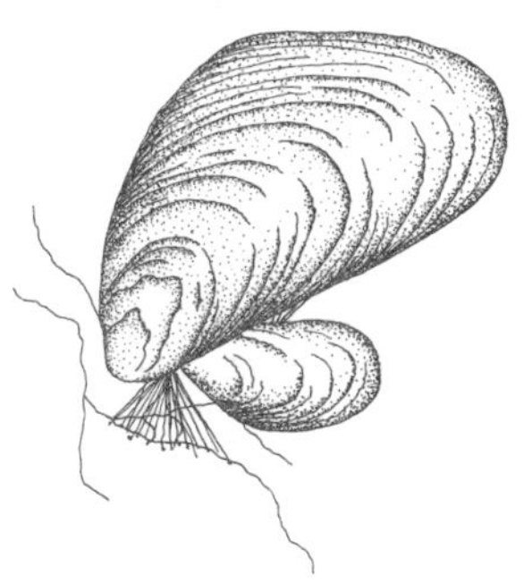

111. *Mytilus edulis* (~5cm long)

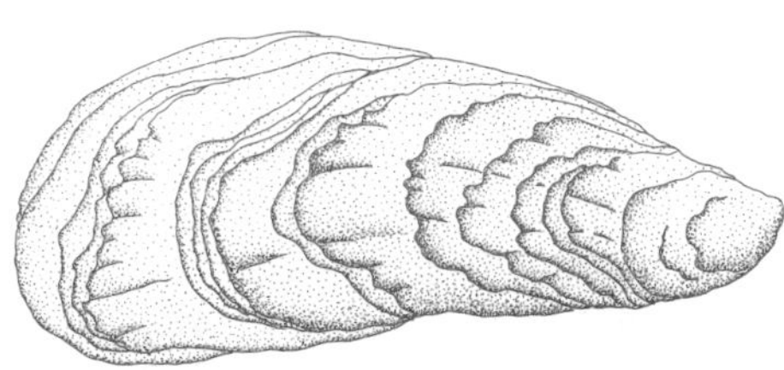

112. *Crassostrea gigas* (~15cm long)

113. *Hinnites giganteus* (~10cm diameter)

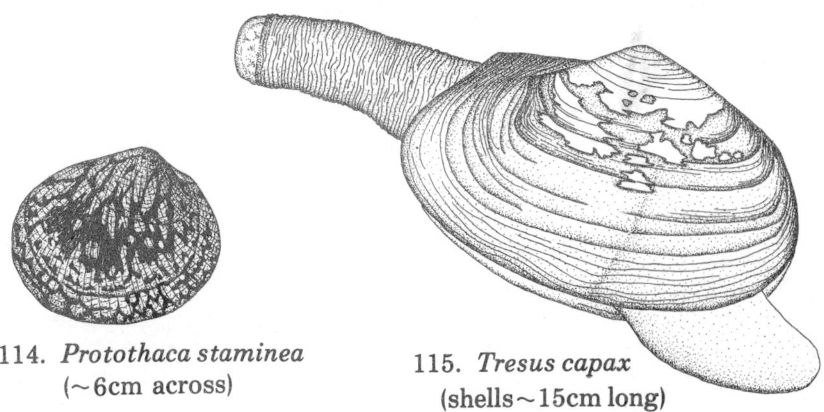

114. *Protothaca staminea* (~6cm across)

115. *Tresus capax* (shells~15cm long)

fishing license.) Mother Nature as well as the authors will rest easier for your efforts in this regard.

Mytilus edulis is generally smaller and smoother-shelled than its rocky-coast relative. This species is abundant in Humboldt Bay and similar areas all along the Pacific coast and may be found clustered on pilings or floats or rock surfaces. The bay mussel is also edible and is grown commercially in some parts of the world. In areas of mixed or intermediate surf, the two species of *Mytilus* may occasionally be found growing together. But unless the water is calm, *M. californianus* is competitively dominant.

Oysters are sessile bivalves which cement themselves to hard substrates by one shell after settling from the larval stage. *Crassostrea gigas* (fig. 112), the Japanese oyster, is the only species likely to be encountered locally. These animals are not native to California shores, but are imported as young, newly settled individuals (called **spat**) from Japan and raised to market size in appropriate habitats such as Humboldt Bay; thus, one will not find them except in commercial beds. Their reproductive success here is poor, and thus importation of the young is a continual process. One Pacific coast firm south of here now raises and sells several species of young oysters in hatcheries. The animals are shipped around the world to mariculturists and the like for subsequent rearing. This organization has a mechanism for preventing the larvae from attaching to a substrate, thus orders are received as bags of loose baby oysters each two to three millimeters long, and the techniques of rearing and handling are much less expensive than with oysters which are attached to one another or old shells. *Crassostrea virginica*, the American east coast oyster, is also imported and reared in some bays further south. *Ostrea lurida*, the small but tasty Pacific or Olympic oyster, occurs in natural populations north of here and is a commercially important species in Puget Sound.

We have mentioned earlier the phenomenon of the male-to-female sex reversal in certain invertebrates. Some oysters carry this odd habit a step further in that they alternate from male to female to male to female . . . and so on. Such goings on are viewed by Jerome Tichenor as follows:

> "Consider the case of the oyster,
> Which passes its time in the moisture:
> Of sex alternate,
> It chases no mate,
> But lives in a self-contained cloister."
> (pg. 31, **Poems in Contempt of Progress**,
> Boxwood Press, 1974)

The rock scallop, *Hinnites giganteus* (fig. 113), is another bivalve which cements itself to solid objects by one shell. This species usually occurs subtidally but may be seen on floats and dock tires in Humboldt Bay. *Hinnites* attaches to the substrate when it reaches a size of 2-3cm. Until that time it swims about near the bottom by clapping the shells together and forcing water out in one direction. The lower shell then grows to conform to the surface of the substrate, and the upper shell becomes rough and often heavily encrusted with other organisms such as sponges and bryozoans.

Protothaca staminea (fig. 114) is commonly known as the rock cockle or the little-neck clam (because of its short siphons) and is among the favored edible bivalves. It is especially tasty in chowder. Look for *Protothaca* in stable substrates such as hardpacked sand and gravel in Humboldt Bay or in the coarse sediments under rocks along semi-protected coasts. These clams are weak burrowers and cannot survive in areas where the sediments are subjected to the direct force of waves. The short siphons restrict this clam to the upper few centimeters of the substrate, and they may often be encountered at or near the surface beneath stones. Some clam diggers have discovered the ease with which *Protothaca* may be collected in such boulder rubble areas as the rocky coast between Trinidad and Clam Beach. Unfortunately a few of these people have utilized collecting methods which are destructive to the intertidal environment. The northern author reports some major damage in these areas where high numbers of rocks have been left overturned and the underlying sediments raked and shoveled. The simple practice of replacing rocks in their original positions would have prevented much of this disastrous situation. We wish people would be more concerned and more careful.

Gaper or horseneck clams belong to the genus *Tresus* (once known as *Schizothaerus*). *Tresus capax* (fig. 115) is by far the more common of the two species found in Humboldt Bay. The other, *T. nuttallii*, predominates in bays to the south such as Tomales Bay. The latter species may be recognized by the presence of two fingernail-like plates at the end of the siphons. *T. capax* lacks these plates. *Tresus* is the largest type of clam you are likely to encounter along the north coast, often exceeding 15cm in shell length. The siphons are long and fused together into the fleshy extension referred to as the "neck." The siphons are contractile but because of their size cannot be pulled completely inside the shells. Because of this feature and the overall large size of the clam's soft parts in general, the shells do not completely close at either end, but rather "gape." Both species of *Tresus* commonly house other animals within their mantle cavities, the most obvious being "pea crabs" of the genus *Pinnixa*. At least one crab of each sex will usually be found in nearly every *Tresus* examined. The male crabs are smaller and more slender than the plump females. The large commensal nemertean *Malacobdella grossa* may also be found while poking around inside *Tresus*. Although *Tresus* is a weak burrower, it is difficult to extract from the mud or sand as it may occur at depths of over .5m. The siphons may be seen to pull down from the surface of the substrate when disturbed, but the clam will stay put.

Except for its more rounded shells and smaller size, *Mya arenaria* (fig. 116) somewhat resembles *Tresus*. This animal, introduced accidentally from the east coast, is commonly called the soft-shelled clam. *Mya* is generally found in slightly brackish water near the mouths of rivers and sloughs. A tasty clam, *Mya* is of considerable commercial importance especially on the Atlantic coast where it occurs in extremely high numbers.

Perhaps the most highly prized of the edible clams is *Siliqua patula*, the razor clam (fig. 117). The northern end of Clam Beach, near the mouth of Strawberry Creek, is the most popular area for hunting razors along our north coast, and that small section of beach receives tremendous pressure during the appropriate seasons. One may encounter literally hundreds of avid clam diggers there during early-morning summer low tides, each hoping to capture their limit of razor clams. To do so is not always an easy task, for *Siliqua* has a strong muscular foot and a streamlined shape which allow it to burrow rapidly, and one must be quick with a shovel to prevent them from digging too deep for recovery. When placed on the sand surface, these clams can completely bury themselves within a few seconds. It is because of this burrowing ability that razors are able to survive in the unstable sandy environment of regions such as Clam Beach and despite the attempted predation by a great many humans.

Among the most easily seen clams in bays is the heart cockle, *Clinocardium nuttallii* (fig. 118). This robust, thick-shelled clam commonly reaches seven or eight centimeters across the shells, and tends towards an inflated globose shape. The very heavy radial ridges on the shells distinguish *Clinocardium* from other clams of similar size and habitats. Because of the extremely short siphons, this species is restricted to burrowing close to the surface of the substrate and in fact often simply lies on top of the mud. Also, the globose shape is not designed for easy movement through the mud or sand. Heart cockles are thus quite exposed to predators and depend upon means other than burrowing for protection. The thick shells aid to some extent in this respect, and they are also known to display a rather dramatic escape response when predators such as seastars (especially *Pycnopodia*, (fig. 126) threaten. This reaction involves the use of the large, L-shaped foot which may be extended from between the shells and employed as a vaulting pole which by straightening suddenly forces the animal to leap off the mud and somersault off in a manner quite unlikely and spectacular in a sedentary type of beast.

The genus *Macoma* (fig. 119) includes relatively small, somewhat flattened clams common in the muddy and fine sandy areas of protected bays. The chalky white shells often show patches of the black, parchment-like **periostracum** on their surfaces. The periostracum is an organic material formed on the outside of a variety of mollusks' shells. The siphons of *Macoma* are long, thin, and separate from one another. The incurrent siphon is extended out of the substrate and bent over toward it where it functions like a vacuum cleaner, sucking up the film of organic material from the surface of the mud. The exhalent siphon in most species expels water above the substrate; in the bent-nosed clam *Macoma nasuta*, the excurrent siphon is held beneath the substrate. The latter species may pulse water out through the incurrent siphon in a sort of backflushing action. In certain parts of the bay habitat, *Macoma* may occur in such numbers that a single shovelful of sedi-

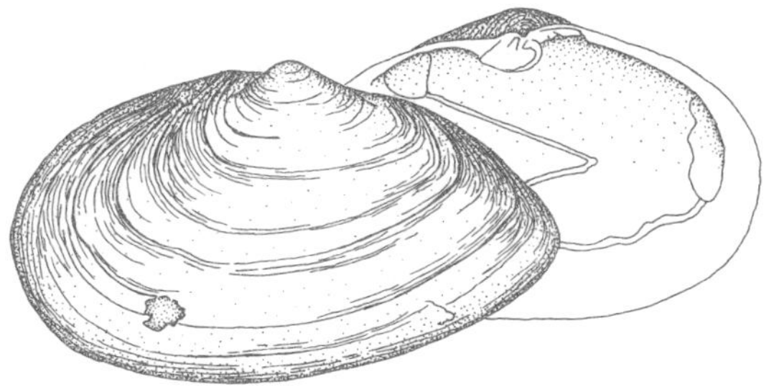

116. *Mya arenaria* (~8cm long)

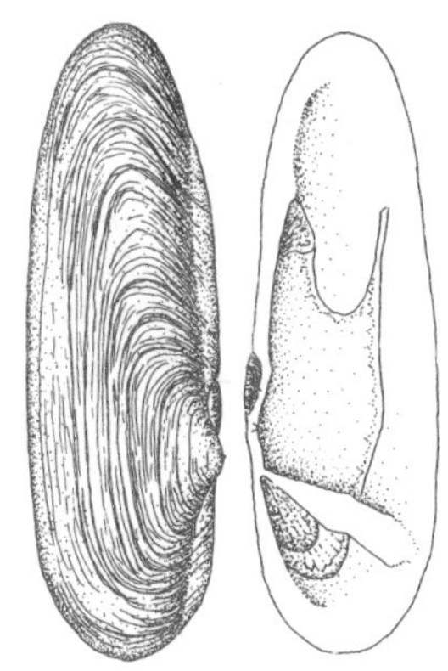

117. *Siliqua patula* (~10cm long)

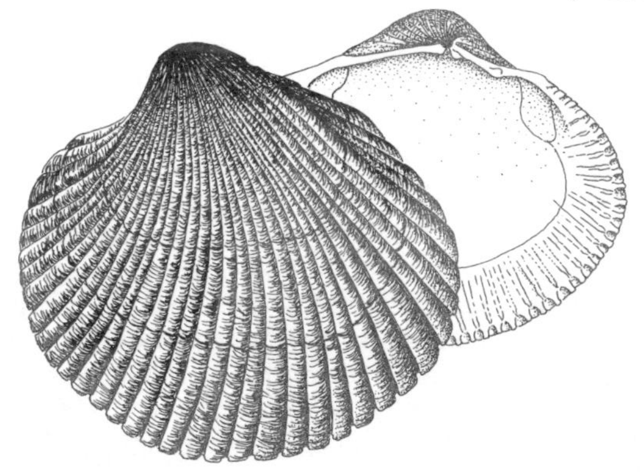

118. *Clinocardium nuttallii* (~7cm across)

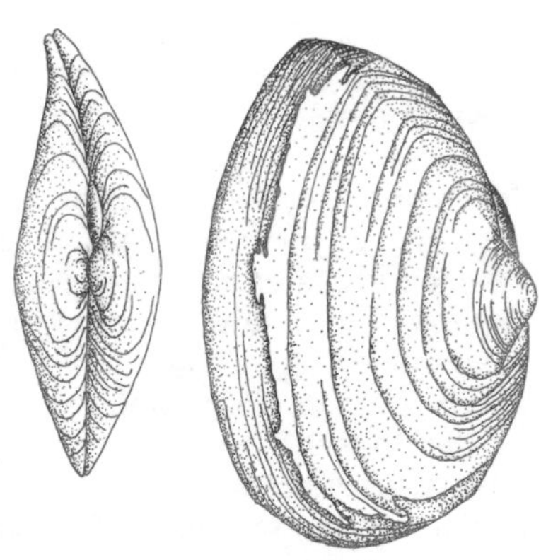

119. *Macoma nasuta* (~5cm long)

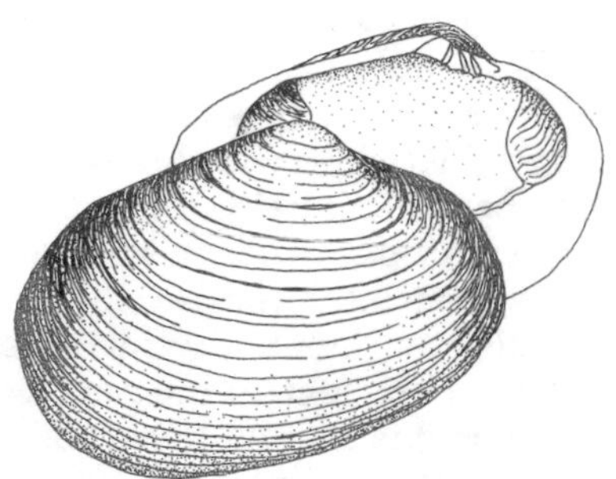

120. *Saxidomus nuttalli* (~12cm long)

ment may turn up a dozen or so individuals.

One of the largest bivalves in Humboldt Bay is the Washington clam, *Saxidomus nuttalli* (fig. 120). It is, however, not as abundant as several of the species mentioned above. Specimens over 15cm in shell length are not unusual. Well-developed concentric growth lines are particularly obvious in this species. The siphons are long and fused, as in *Tresus,* but may be pulled within the shells. Washington clams generally occupy habitats similar to those of gaper clams.

Class Cephalopoda

the octopuses and squids

The octopuses and squids have given up the sedentary life-styles of most other mollusks and have become hunting predators capable of rapid swimming. With the exception of the tropical chambered nautilus, all living cephalopods have either lost their shell or have moved it inside the body and greatly reduced its size and weight. The nervous system is heavily concentrated as a large, well-developed brain associated with the complex sense organs necessary for an active, prey-seeking mode of life. Most obvious among the sense organs are the large and complex eyes, of similar construction to vertebrate eyes, which permit cephalopods a high degree of visual discrimination. The circulatory system has become virtually closed and more efficient than the open type of other mollusks. Accessory blood-pumping organs are present to maintain blood pressure, especially in the large gills suspended in the mantle cavity.

The feeding structures are also highly modified from the presumed ancestral type. The radula is often used, in octopuses, as a drill to bore tiny holes through the shells of more conventional mollusks. Certain salivary glands have become poison-secreting organs for paralyzing or killing prey. Cephalopods possess a horny beak at the entrance to the mouth, surrounded by eight (in octopuses) or ten (in squids) suckered arms. The prey is captured by the arms, bitten by the beak, and immobilized with the poison.

The remarkable and extremely rapid color changes displayed by cephalopods are brought about by the activity of pigment-containing cells called **chromatophores,** which lie beneath the skin. These cells may be differentially contracted and expanded to create a variety of colors and patterns. In addition, these animals, especially the octopuses, can alter the texture of their body surface from smooth to wrinkled or bumpy. While these modifications of appearance are generally thought to function as camouflage, they are also used when the animal is excited or alarmed, and during certain mating activities and territorial displays.

In addition to the adaptive qualities mentioned above, cephalopods also possess an ink sac which is capable of releasing a cloud of pigment granules suspended in mucus into the water. The pigment is **melanin,** the same deep-brown or nearly black substance responsible for tanning in humans. The dark cloud tends to hang together for a short time and acts as a decoy allowing the cephalopod a chance to escape predation. Some zoologists suggest that certain chemicals emitted with the pigment may actually repel predators and/or act to desensitize their organs of smell.

Squids are fully pelagic creatures and are not encountered intertidally. The common bait or food squid of our coast is *Loligo opalescens,* frequently picked up by local trawl fishermen out of Eureka. There is an active fishery for *Loligo* in Monterey, and these creatures appear frequently on tables in the Italian communities along that portion of the California coast. The authors' mother used to prepare squids in a traditional Sicilian, "calamari" manner; simmered in a spicy red sauce, with wine... but we digress. Most good cookbooks contain this and other less imaginative, but adequate, recipes.

The squids include the largest living invertebrates, the giant deep-sea *Architeuthis.* Specimens exceeding 15 meters in length, including the arms, have been recorded from the north Atlantic.

Small octopuses (probably *Octopus rubescens*) are not uncommon in the low intertidal zones of the northern California coast, but are infrequently seen because of their camouflage and secretive habits. We have seen them in areas of boulders protected from the direct force of waves, in the surf grass beds near Patrick's Point, and in the eelgrass regions in Humboldt Bay. It is somewhat difficult to go out and intentionally find an octopus; rather, they are generally located almost by accident when searching among likely habitats for other creatures. This intertidal octopus rarely exceeds 30cm between the tips of the extended arms, and most individuals are significantly smaller. It can, however, inflict a very painful, and occasionally serious, bite with its beak. Although these creatures are normally docile and want only to be left alone, if handled roughly they will press the mouth area against the skin of their tormentor and defend themselves with a quick nip. Beyond the possibility of infection from the wound, some poison is generally introduced at the time of biting. The effect of the poison depends upon the amount received and the general sensitivity of the victim, ranging from a bee-sting-like swelling and irritation to nausea or perhaps worse in extreme cases.

Octopus dofleini is among the largest of all the world's octopuses, and occurs subtidally along our coast. The northern author has seen three or four stranded among intertidal pools at low tides, but such occurrences are rare. These creatures may measure several meters between arm tips and possess basket-ball-sized bodies. They are often captured in the crab traps of local fishermen, as crabs are among this creature's favorite foods. The Humboldt State University Marine Laboratory at Trinidad often has one or more of these large cephalopods on display.

THE LOPHOPHORATES (crest or tuft bearers)

phoronids, brachiopods, and ectoprocts

Three phyla of coelomate animals are commonly lumped together under the category of "lophophorates." The **lophophore** itself is a ring of hollow ciliated tentacles which surround the mouth and function in the acquisition of food (and also gas exchange). The coelom extends into these tentacles and the coelomic fluids are used as a hydraulic skeletal system to hold the tentacles erect. Whether the lophophorates actually constitute a natural grouping of animals or not is debatable, but they are traditionally considered to do so.

Phylum Phoronida (nest bearers)

About 15 species of **vermiform** (worm-shaped) lophophorates are included in the small phylum Phoronida, but only one species (*Phoronopsis viridis,* fig. 121) is likely to be found along our shores. These animals often occur in very high numbers in the mud-sand substrates of bay tidal flats. The worms are about 7-10cm long and live in sandy tubes 4-5mm in diameter and perhaps twice as long as the worm's body. The tubes are set vertically in the substrate with no more than a few millimeters projecting above the surface. The tiny green lophophores of *Phoronopsis* may be seen extending out of the tubes when covered by water. In areas of high densities, dozens of individuals may occur in a 10 x 10 cm area. *Phoronopsis* is a filter-feeder, setting up water currents with the ciliated surfaces of the lophophoral tentacles. The digestive tube is U-shaped, the anus being placed near the mouth but outside of the ring of tentacles. These unsegmented worms possess a complicated, closed circulatory system. As is the case with so many near-shore benthic animals, phoronids undergo indirect development, producing a free-swimming larva which eventually settles to burrow and form its tube as a juvenile worm.

Phylum Brachiopoda (arm-feet)

the lamp shells

The lamp shells are locally represented intertidally by *Terebratalia transversa* (fig. 122), the only brachiopod likely to be encountered here. These brachiopods are not common on the shore but may occasionally be seen attached to the protected sides of low-intertidal rocks (they are very abundant subtidally). When first encountered, they may be mistaken for some bizarre clam, because the body is encased within a pair of hard calcareous shells similar to those of bivalved mollusks. These

121. *Phoronopsis viridis*
(worm ~6cm long)

122. *Terebratalia transversa*
(~3cm across)

creatures are attached to the rock surface by a fleshy stalk called the **pedicle** which originates near the hinge of the two valves or shells, unlike the means of attachment of any clam. Internally they are nothing at all like clams; in fact the shells are oriented dorso-ventrally (top and bottom) relative to the brachiopod's body, rather than laterally (on the sides) as in bivalves. The lophophore is between the shells and consists of a pair of coiled ridges bearing the tentacles which the animals use to create their feeding currents. They produce free larval stages which undergo a rather drastic metamorphosis upon settling. If you happen to find a specimen of *Terebratalia*, we hope that you will observe it in place, and not break it from the rock upon which it lives, for once removed, they cannot reattach. One nice population of these animals has nearly disappeared from the low rocks south of Trinidad Head, probably at the hands of indiscriminate collectors.

Phylum Bryozoa (moss animals) = Ectoprocta

Bryozoans are tiny lophophorate creatures which live in colonies of various shapes, colors, and sizes (fig. 123, 124, 125.) Except for those which have truly distinctive growth patterns, bryozoans are extremely difficult to identify, and we make no attempt to do so in this little book. The erect, branching (**arborescent**) forms are often mistaken for algae or colonies of hydroids. Upon close examination (with a hand lens), however, one can detect the tiny (1mm or less) compartments which house the individual animals or **zooids**, and, if the colony is under water and undisturbed, the tiny ring of tentacles extended to feed. The encrusting bryozoans (fig. 124) usually form reddish, yellowish, or whitish calcareous and rough patches on the undersides of rocks or on blades of algae or eelgrass. Again, careful examination will reveal

123. Arborescent bryozoan colony (~2cm high)
inset, magnified portion of colony

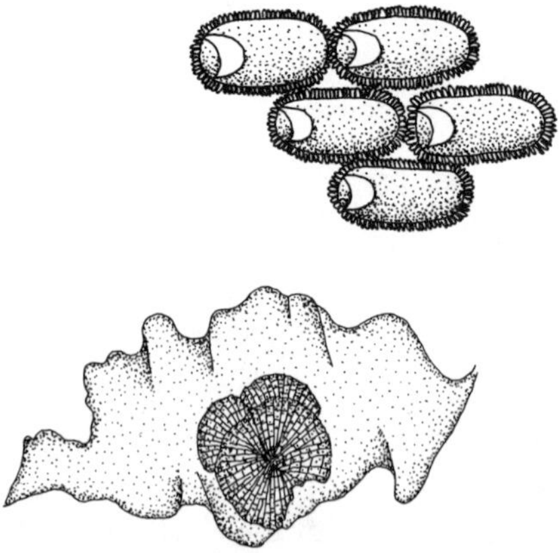

124. Encrusting bryozoan colony (~2cm across)
inset, individual zooids

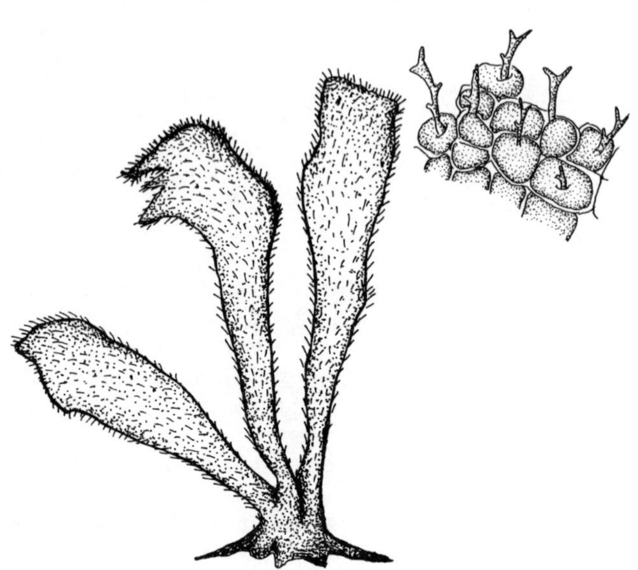

125. *Flustrellidra corniculata* colony (~8cm high)
inset, magnified portion of colony

the net-like pattern on the colony formed by the compartments housing the zooids and prevent misidentifying these animals as sponges or some other encrusting organism. The most common encrusting forms are likely to be of the genus *Membranipora.* One easily identifiable bryozoan is *Flustrellidra corniculata* (fig. 125). This species grows in colonies which appear as rather thick, somewhat soft and fleshy, light brown or tan blades the surfaces of which are covered by minute, dark, antler-like spines (use a hand lens). The dull-brown colonies are often seen stranded on beaches after heavy weather and resemble bits of seaweed. Attached colonies may be found among low intertidal rocks in areas protected from the direct force of waves.

As in the other lophophorates, bryozoans use their tentacles to filter tiny particles of food from the water. Certain types of bryozoans may display drastic **polymorphism** between members of the colony similar to the condition described earlier for most types of Hydrozoa. The most common type of non-feeding zooid is the **avicularium** which resembles a microscopic bird's head and beak and functions in keeping the colony clean and free of debris and settling organisms. The embryogeny of bryozoans involves a free-swimming larval stage which settles and forms a single zooid which gives rise to the colony through asexual budding.

PHYLUM ECHINODERMATA (spiny-skinned)

the seastars, sea urchins, sea cucumbers, etc.

Echinodermata means spiny-skinned, and indeed many members of this phylum display spines of various sorts. This group includes the familiar seastars, sea urchins, and sand dollars, along with some less frequently seen forms such as brittle stars and sea cucumbers. These coelomate animals generally display pentaradial, or five-part, symmetry, and while this body plan may not be immediately obvious in all types, it is certainly evident in such things as the common seastars. This symmetry lends itself well to the sedentary life styles of most echinoderms, allowing them to react to environmental conditions on all sides. Most zoologists agree that the radial symmetry seen in members of this phylum was secondarily derived from some bilateral ancestor and does not indicate a relationship to other radially oriented animals like cnidarians. One piece of evidence for this belief is that echinoderms pass through larval stages which display bilateral form. One group, the sea cucumbers, has begun to revert back to a bilateral condition, and the animals appear as rather fat, worm-like creatures with a front and back end. Echinoderms are oriented on an oral-aboral axis, similar to the cnidarians, but with the oral or mouth side usually directed down against the substrate (exceptions will be noted in the discussions). In the sea cucumbers, the oral side has become the front end and the aboral side the back end. The spines mentioned above are outgrowths of a calcareous internal skeleton which is composed of plates which may be fused together to form a solid case (urchins) or which freely articulate with one another so that the animal is quite flexible (many seastars and brittle stars). In the sea cucumbers, the plates are usually present as isolated bits of skeletal material beneath the skin, so the animal is rather fleshy, although in some species these dermal ossicles may fuse to form a very rigid skeleton.

Echinoderms possess a unique hydraulic system within their bodies called the **water-vascular system**. This system apparently operates on the fluid of the coelom. A complex system of tubes and semipermeable membranes extends through the body and terminates in the characteristic **tube feet** of these creatures. The suction-cup "feet" serve a variety of functions, notably locomotion, attachment, and in some cases feeding. Various muscles and water reservoirs associated with the water-vascular system allow the animal to attach the tube feet securely to any hard substrate. The coordinated movement and attachment and detachment of the feet allow the animal to crawl over surfaces or clamp down tightly against rocks. One need only view the underside of a seastar or sea urchin moving over the side of a glass aquarium to have the coordinated movement of the tube feet made apparent.

Of the five generally recognized classes of Echinodermata living today, four are commonly encountered in local intertidal habitats. These are the classes Asteroidea (seastars), Echinoidea (sea urchins and sand dollars), Ophiuroidea (brittle stars or serpent stars), and Holothuroidea (sea cucumbers). The fifth class is the Crinoidea and includes the rather bizarre and presumably primitive subtidal and deep-water sea lilies and feather stars. The reader may encounter some different classification schemes in recent texts, but the above traditional arrangement is a bit easier to manage.

Class Asteroidea
the seastars

The seastars are certainly the most familiar echinoderms and among the most characteristic animals of the world's sea shores, practically symbolizing seashore life. Most of these animals are carnivores and feed on a variety of other invertebrates. A few, such as *Henricia* (fig. 133), are **suspension feeders**, eating bits of detritus stirred up from the surface of the sediment. Bivalved mollusks are a favorite food for many of the local seastars; the predation on *Mytilus* by the ochre star (*Pisaster ochraceus*, fig. 128) has already been discussed. Asteroids manage to eat bivalves by wrapping their arms around the prey with the mouth pressed tightly against the opening between the shells. They possess an eversible stomach which can be extended out of their body and in between the clam's shells. So flexible is this stomach that it may be inserted through tiny gaps of a millimeter or so even when the bivalve is closed as tightly as possible. In many instances a seastar will exert pull on each shell with its tube feet which are located in the **ambulacral grooves** along the underside of each arm. This action eventually results in the clam slightly opening its shells allowing easier insertion of the seastar's stomach. In any case, the eversible stomach brings with it powerful enzymes to begin the digestion of the prey even before it is taken within the body of the seastar. As the powerful muscles of the clam are acted upon by these enzymes, the valves gape even more allowing further insertion of the predator's stomach. Although many animals are attacked and eaten by seastars, the versatile stomach may be employed to digest and engulf dead animal matter as well. the material is partially digested by the protruded stomach and the resulting "soup" is then taken in as the stomach is withdrawn.

Members of the class Asteroidea may be found in nearly every type of coastal habitat in northern California except wave-swept sandy beaches. Individual species, however, are generally restricted to particular substrates; thus, where the animal is found is often a clue which will aid in identification as we have indicated for most of the previous groups. Most of the common species are included in the key below.

Key to the Common Seastars of Northern California

1. With many more than 6 arms 2
1. With 5 or 6 arms 4
2. Body soft, "mushy"; arms very flexible; usually 15 or more arms; animal may be .5m or more in diameter .... *Pycnopodia helianthoides* (fig. 126)
2. Body somewhat crusty, more rigid; arms rather stiff; usually 9-13 arms.. 3
3. Upper surface of arms with a pale to dark purplish line extending from the central disc to the arm tip .... *Solaster stimpsoni*
3. Upper surface of arms more or less uniformly yellowish to orange. .... *Solaster dawsoni* (fig. 127)
4. Small, usually less than 5cm diameter; whitish-grey; with 6 arms. .... *Lepasterias* (fig. 132)
4. Usually larger than 5cm diameter; color variable; typically with 5 arms (an occasional specimen may be found with an "extra" arm, the result of odd growth or injury, and subsequent regeneration of an additional part)................................ 5
5. Arms broad, nearly webbed at their attachment to the central disc (fig. 131) 6
5. Arms not distinctly webbed.................................. 7
6. Surface somewhat rough or crusty; color orange, occasionally with purple blotches .... *Patiria miniata*
6. Surface smooth and slick; color mottled, dull orange-red and grey. .... *Dermasterias imbricata* (fig. 131)
7. Surface with small bumps but no obvious spines; color bright red

or reddish-orange; arms thin and nearly round in cross section; animal small, usually less than 10cm diameter .... *Henricia leviuscula* (fig. 133)

7. Surface with distinct spines; color variable; arms not as above; animal larger than 10cm in diameter 8
8. Color uniformly pink.............. *Pisaster brevispinus* (fig. 129)
8. Color not uniformly pink.......................................9
9. Arms relatively thin, 4-5 times as long as the diameter of the central disc .... *Evasterias troschelii* (fig. 130)
9. Arms thicker, only 2-3 times as long as the diameter of the central disc.............................. *Pisaster ochraceus* (fig. 126)

Of the three species of "many-rayed" stars, *Pycnopodia helianthoides* (fig. 126), is by far the most common along our shores. The body of this rather flexible, somewhat fleshy animal commonly exceeds 30-40 cm in diameter (the largest seastar on our coast), although smaller individuals are more numerous. They occur in a variety of colors, including yellow, reddish-brown, and pale lavender. *Pycnopodia* is an extremely fast crawler as seastars go, and you can easily see them progress in a gliding motion over the substrate, propelled by the walking action of thousands of tube feet. This species is an active predator on a variety of invertebrates, especially mollusks. Look for *Pycnopodia* in low intertidal zones of semiprotected rocky areas and around sand flats and eelgrass beds in bays. These seastars are not very resistent to drying and will rarely be encountered out of water.

Solaster dawsoni (fig. 127) and *S. stimpsoni*, the sun stars, are characterized by the possession of about 10-12 arms. They range along the northern California coast and throughout much of the Pacific Northwest, yet while *S. stimpsoni* may be seen intertidally, *S. dawsoni* prefers somewhat deeper water. The former species feeds largely on sea cucumbers, while the latter, oddly enough, eats the former.

The most frequently encountered seastar in California is *Pisaster ochraceus,* the common ochre star (fig. 128). This animal is usually associated with mussel beds in which it feeds upon the mussels themselves and small barnacles. This durable animal is one of the few local asteroids capable of withstanding the direct force of large waves and extended periods of exposure to air at low tides. *Pisaster ochraceus* displays a multitude of body colors (red, yellow, brown, orange, purple, etc.). No one knows what the adaptive significance of the different colors is. At low tides the animals are generally clustered just below the mussel zone, moving upwards to feed when the tide comes in. The ochre star is not restricted to rocky shores and may be frequently encountered on pilings and other hard substrates in bays wherever suitable food sources are available.

Pisaster brevispinus (fig. 129), a close relative of the ochre star, is characterized by its consistent pale-pink coloration. This species is adapted to living on softer substrates than is *P. ochraceus* and is common in Humboldt Bay and along the sandy coves between rocky

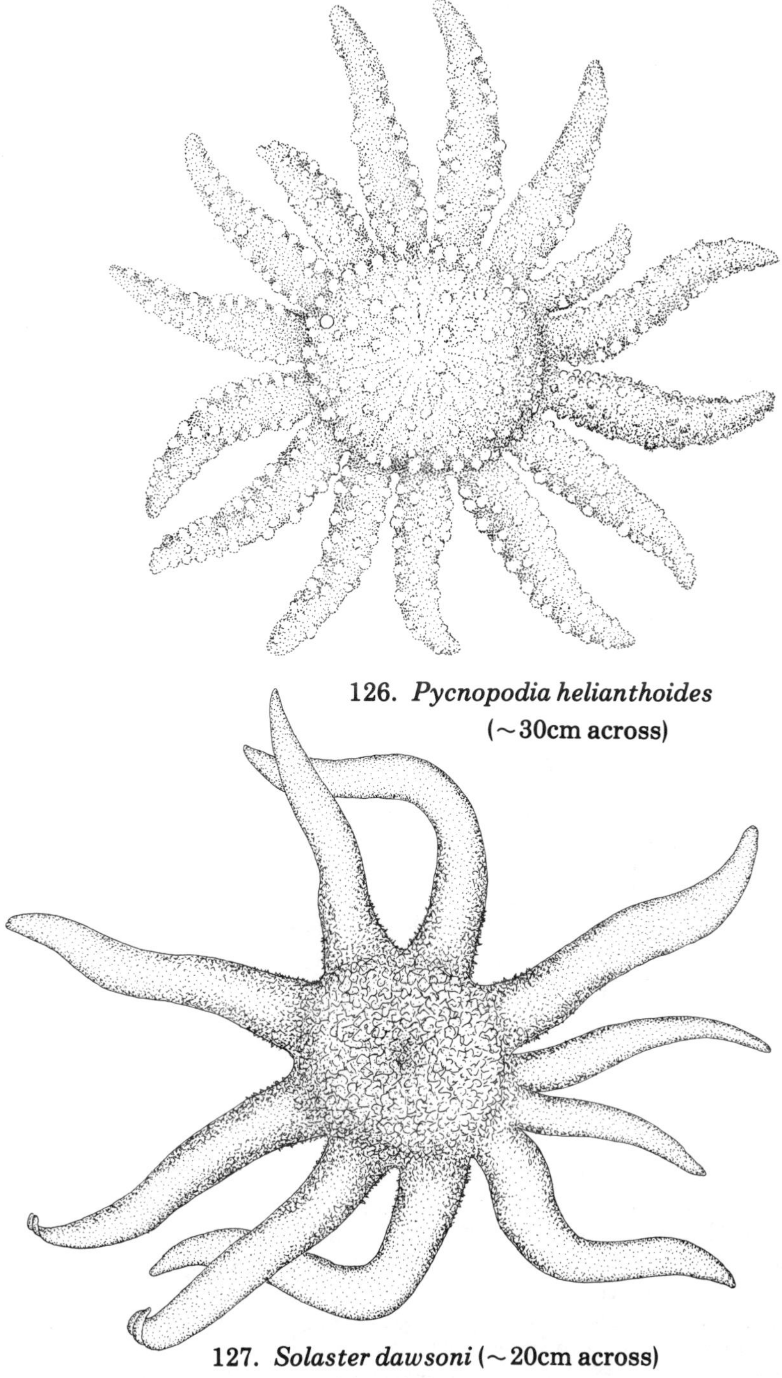

126. *Pycnopodia helianthoides* (~30cm across)

127. *Solaster dawsoni* (~20cm across)

points. *Pisaster brevispinus* preys on snails and small clams.

Evasterias troschelii (fig. 130) is often mistaken for a *Pisaster* because of their similar size and variable coloration. The thinner and relatively longer arms of *Evasterias* distinguish it from *Pisaster*. *Evasterias* is reported to be more abundant subtidally, but it is not uncommon among low intertidal rocks of the north coast where it feeds largely on clams and mussels.

Patiria miniata and *Dermasterias imbricata* (fig. 131) are called the bat star and leather star respectively. They are separable on the traits given in the key above, but both are characterized by the webbed appearance between the arms where they join the central disc. *Patiria* is uncommon along the extreme northern California coast, but it is very abundant further south. *Dermasterias*, on the other hand, is extremely abundant in protected mid- to low-intertidal rocky pools all along the north coast. In some areas around Trinidad and Luffenholtz Beach (especially near Cimberg's Rock), we have seen dozens of these animals at a glance. *Dermasterias* is also common on hard substrates in Humboldt Bay. Neither of these species can cling very tightly to the substrate, and thus they avoid areas of heavy wave action.

There are two relatively small, but common, seastars likely to be seen here. *Leptasterias* (fig. 132) is the only local asteroid which consistently bears six arms. Most individuals are less than 3cm in diameter, and greyish-white in color. *Leptasterias* often occurs in very high numbers among rocks, algae, and surf grass around the zero tide level on rocky shores. Close inspection of the plants in any protected large pool during the summer will reveal hundreds of these animals. *Henricia leviuscula* (fig. 133), the blood star, may reach 15cm between the arm tips, but most individuals are from 4 to 5cm across. *Henricia* feeds on small bits of detritus which become trapped in mucus covering the body. The mucus and food are carried to the mouth by ciliary action and ingested. This animal is perhaps the most striking seastar on our shores; its bright orange to blood-red color makes it stand out against the background of green or dull brown. It inhabits the same general areas as *Leptasterias* and is particularly abundant (or noticeable?) on surf grass.

Class Echinoidea

the sea urchins and sand dollars

Sea urchins and sand dollars are frequently seen dried and discolored on the shelves of curio shops and tourist traps along the California coast. We think that you will enjoy them more at the seashore, alive and in their natural surroundings. All members of this class have skeletons of fused plates, forming a rigid casing, or test, around the internal organs. Such an arrangement does not immobilize these animals, however, for the skeleton is equipped with movable spines and pores through which the tube feet extend. These body extensions both aid in

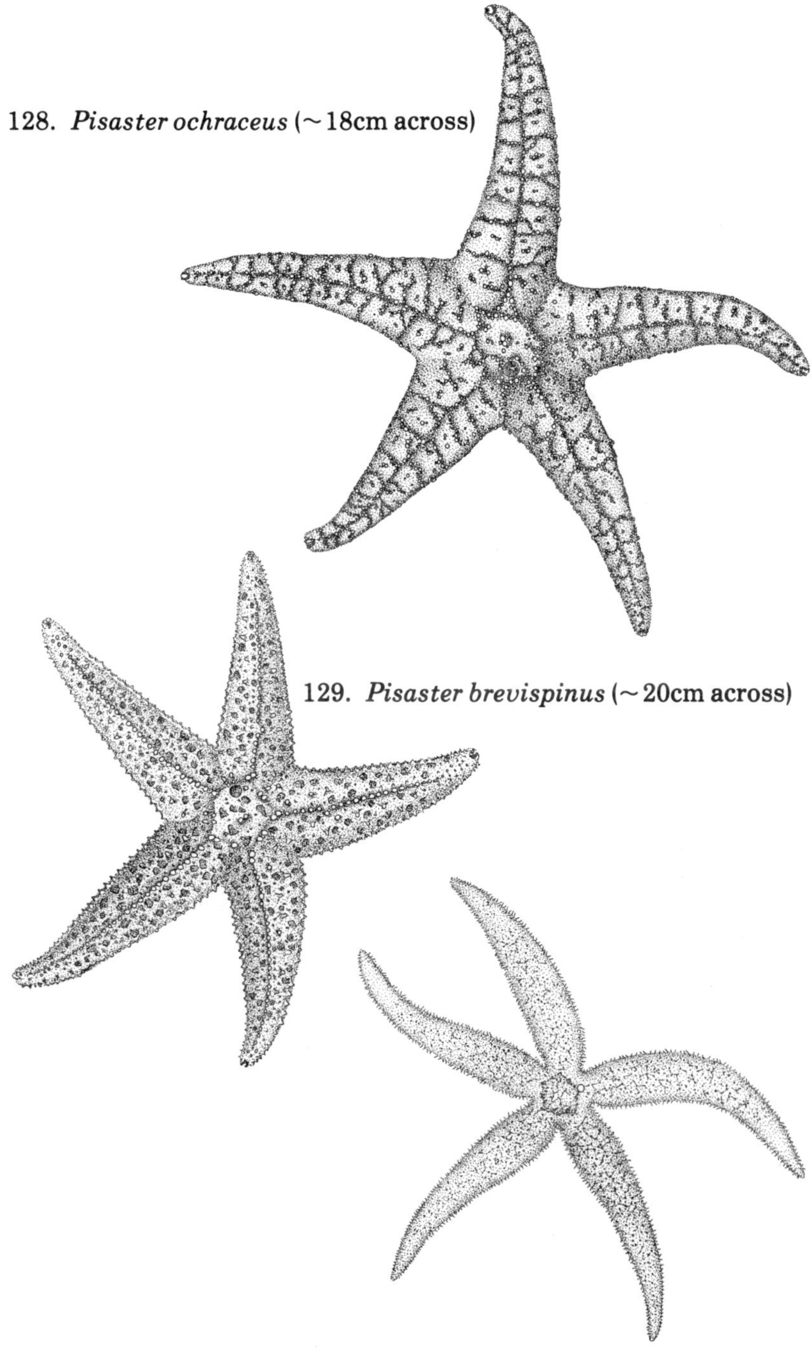

128. *Pisaster ochraceus* (~18cm across)

129. *Pisaster brevispinus* (~20cm across)

130. *Evasterias troschelii* (~15cm across)

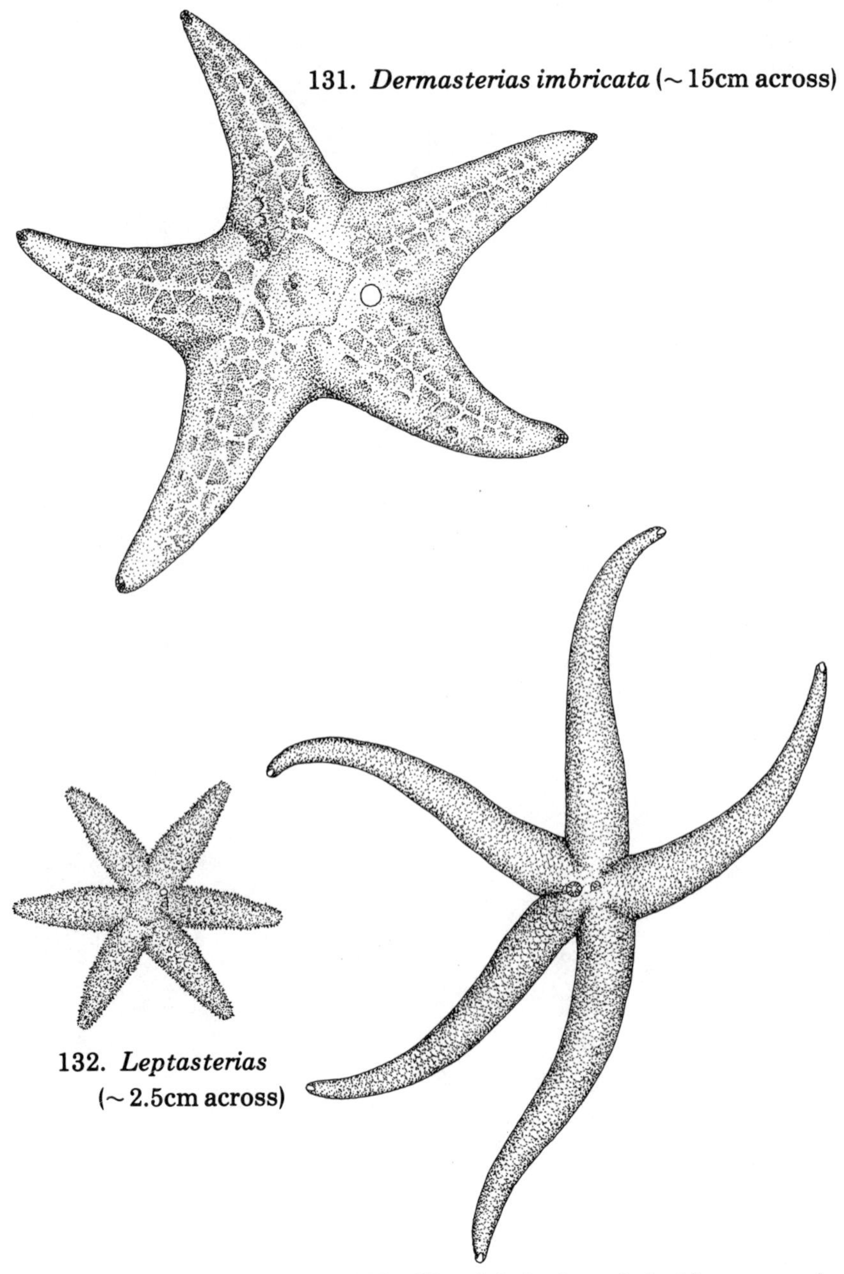

131. *Dermasterias imbricata* (~ 15cm across)

132. *Leptasterias* (~ 2.5cm across)

133. *Henricia leviuscula* (~ 10cm across)

134. *Strongylocentrotus purpuratus* (~ 7cm diameter)

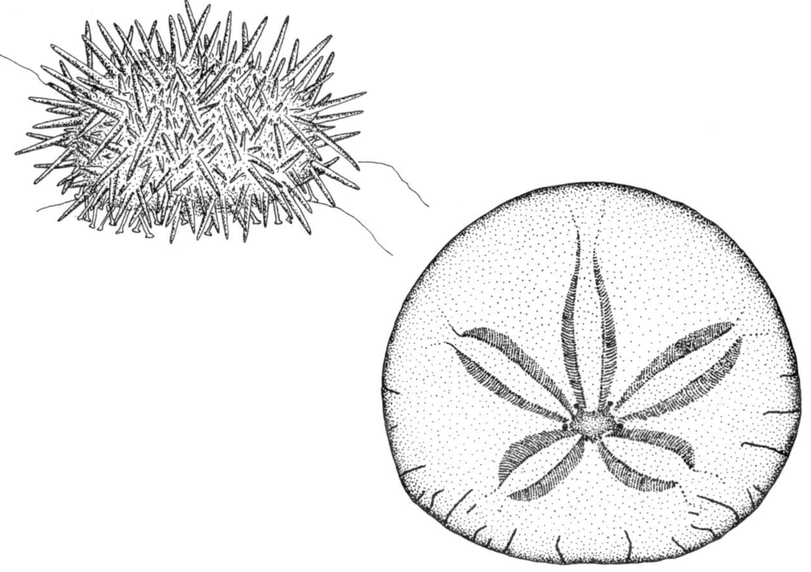

135. *Dendraster excentricus* (~ 8cm across)

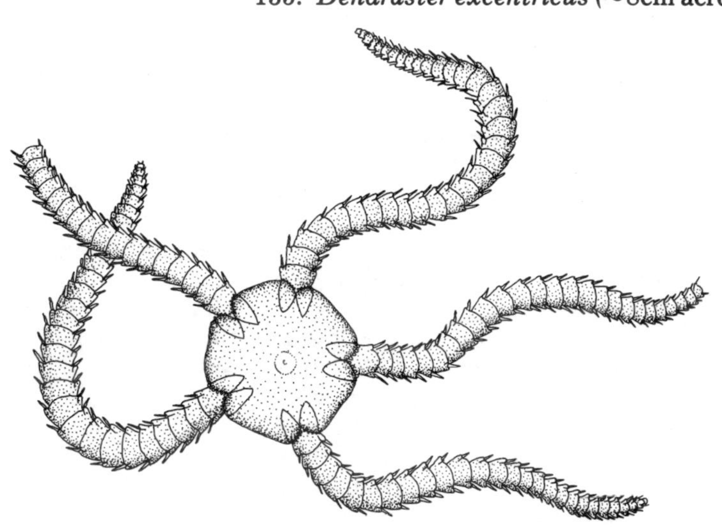

136. brittle star (~ 2.5cm across)

locomotion. The tube feet are arranged in the standard five rows on urchins, but these are not immediately apparent in living animals because the body is covered with spines. The five-part petal design on the upper sides of sand dollars is also a reflection of the pentaradial symmetry seen throughout the phylum. This design is formed by holes in the skeleton through which specialized tube feet protrude which serve as gills.

There are two relatively common species of sea urchins along rocky shores of northern California. The most abundant of these is the purple sea urchin, *Strongylocentrotus purpuratus* (fig. 134). Members of this species are distinctly purple, the spines somewhat lighter in color than the body itself which rarely exceeds 8cm in diameter. They generally occur in large and fairly dense populations among large boulders in the lower intertidal zones on exposed coasts. There are fewer of these animals along the far north coast than prior to the "Big Flood" of 1964, but substantial numbers may be found in appropriate areas north of Trinidad, especially at Abalone Beach. They are also common further south around Shelter Cove. Under typical conditions, *S. purpuratus* feeds by scraping algae off rocks or chewing at large pieces of seaweed with its five teeth located just inside the mouth. The urchins may, however, form excavations in soft stone by the grinding action of the spines. They take up residence in these burrows, live, grow, and enlarge the cavity around and below them, and eventually become too large to get through the opening of their self-formed prisons. Thus trapped, the urchins resort to suspension feeding by extracting food from the water and to catching drifting algae with their tube feet and spines. You may see rock faces that are pitted with empty depressions once occupied by such a population of purple urchins. Young *S. purpuratus* are greenish in color and are often found singly under rocks in slightly higher tidal zones than the rest of the population. These individuals may be easily confused with the green urchin, *S. droebachiensis,* in Washington where their ranges overlap.

Within the very low tidal zones around Abalone Beach, in areas rarely uncovered by the tides, you may find the giant red urchin, *Strongylocentrotus franciscanus.* These great animals are much larger than the purple urchins, often exceeding 15cm in diameter. They range in color from red to very dark purple, nearly black.

The beginning beachcomber may be somewhat reluctant to handle sea urchins, having been made hesitant by tales of poisonous spines and the like. While several tropical species do possess poison associated with their spines or **pedicellariae** (tiny jaw-like structures on the body, common to many echinoderms, and used to keep the body clean), none of the local species is poisonous. The spines are sharp and brittle, however, and can easily penetrate the skin if fallen or stepped upon. The spines are covered with skin and mucus, and a resulting puncture wound can become infected and should be treated accordingly. Spines which break and are left imbedded under the victim's skin may be dissolved with vinegar if the wound is open, or may require some professional medical attention.

The only sand dollar along our coast is *Dendraster excentricus* (fig. 135). This species occurs in very high numbers subtidally on sandy bottoms, although their tests (dead shells) may be found washed up on the shore. The patterns of holes and marks on these empty skeletons reveal something of how the live animals are put together. The oral surface bears two relatively large openings, the central one for the mouth and the peripheral one for the anus. The five-part petal design, as mentioned earlier, indicates the location of special tube feet which serve as gills. The body of a living sand dollar is covered with thin, movable spines which function in locomotion and burrowing. The tube feet are primarily used as touch-sensitive structures and in feeding. Suspended food particles, especially diatoms, are trapped in mucus among the spines and carried to the mouth by cilia and tube feet which line the food grooves which radiate from the center of the oral surface. Young sand dollars are known to eat selectively heavy sand grains (particularly iron oxides) which they store in special regions of the gut, utilizing the material as ballast to keep themselves from being tossed about or dislodged by currents and surge.

Class Ophiuroidea

the brittle or serpent stars

The ophiuroids (fig. 136) are common but often overlooked members of intertidal communities. Their fragile bodies demand that they remain in protected lower-zone habitats, such as algal holdfasts, root tangles of eelgrass, and under rocks. The five arms are very flexible, as the skeletal plates are not fused to one another and articulate freely. The arms, rather than the tube feet, are used for locomotion. Most local species are small, generally less than 3 or 4 centimeters between arm tips, but some tropical species may be more than 20cm across. Ophiuroids display a number of feeding methods and sometimes a single species will employ more than one technique. Filter-feeding, detritus-feeding, scavenging, and even predation are all known to occur in members of this class. Specific identification of many brittle stars is difficult at best, requiring microscopic examination of the arrangement of skeletal plates, spines, patterns of bumps and nodes, etc., and to make matters worse, our local fauna is in need of taxonomic reexamination. We hope an interested student may one day see fit to present us with a good survey of these animals from the north coast.

Class Holothuroidea
the sea cucumbers

The holothurians are, at first glance, unlikely relatives of the other echinoderms. These worm-like or sausage-shaped creatures are superficially quite different from their kin, their bodies being lengthened in the oral-aboral axis. Most of the skeleton has been lost or reduced so that the body is soft and fleshy. They retain the tube feet, however, often in the traditional five rows along the length of the body (in some species the "upper" or dorsal feet have been lost, and a few burrowing forms have no feet at all). Sea cucumbers do not move around very much. The tube feet help anchor the animal in cracks, under rocks, or on other irregular surfaces. Most sea cucumbers feed by means of a ring of branched oral tentacles which collect small food particles (detritus) and stuff them into the mouth. They have evolved a method of "breathing" unique among echinoderms. Whereas most other members of thie phylum exchange oxygen and carbon dioxide through the skin (of the tube feet and special thin areas called **dermal gills**), the cucumbers pump water in and out of their anus. Gas exchange takes place through a highly branched **respiratory tree** which arises from the wall of the hindgut. In certain parts of the world, small fishes of the genus *Carapus* have found this portion of the cucumber's gut a safe place in which to live, and when not feeding rest securely within the anus of the host's digestive tube. Another odd feature of some holothurians involves a method of avoiding predation. When overly disturbed or threatened, a cucumber may spew out all or part of its digestive tract and masses of sticky tubes, a process called **evisceration.** The eviscerated parts may entangle a potential predator and possibly serve as a decoy allowing the cucumber to escape and eventually regenerate its lost organs.

Eupentacta quinquesemita (fig. 137) is perhaps the most frequently encountered local sea cucumber. This creamy-white animal, a few centimeters long, may be seen by the hundreds on the vertical faces of sheltered rocks during the spring and summer months. It also occurs on floats and pilings in Humboldt Bay. The feeding tentacles are rarely seen when the animal is exposed during low tides or disturbed by handling. All five rows of tube feet are present on *Eupentacta,* revealing its basic pentaradial symmetry.

Among the most striking of all cucumbers is *Cucumaria miniata* (fig. 138). This large (up to 25cm long) plump animal is rusty orange in color, and the bushy tentacles are a brilliant reddish-orange. You may occasionally find small individuals of this species attached to the undersides of moderate-sized stones, but most of the larger ones will be lodged securely in cracks and crevices in the low-tide zones with only their bright tentacles betraying their presence. The tentacles retract quickly when disturbed and the animals are nearly impossible to remove from between the rocks, and you should not try as they are easily damaged.

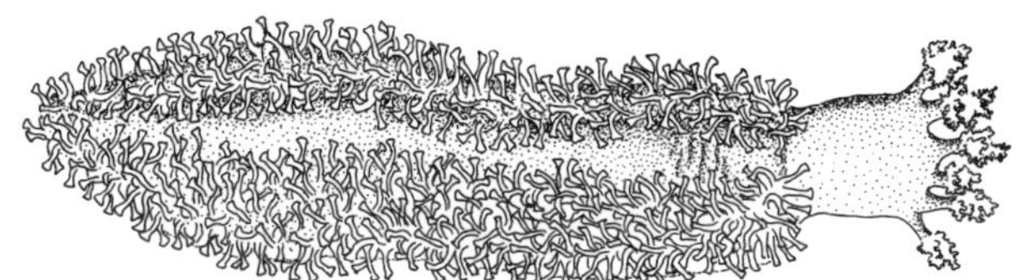

137. *Eupentacta quinquesemita* (~6cm long)

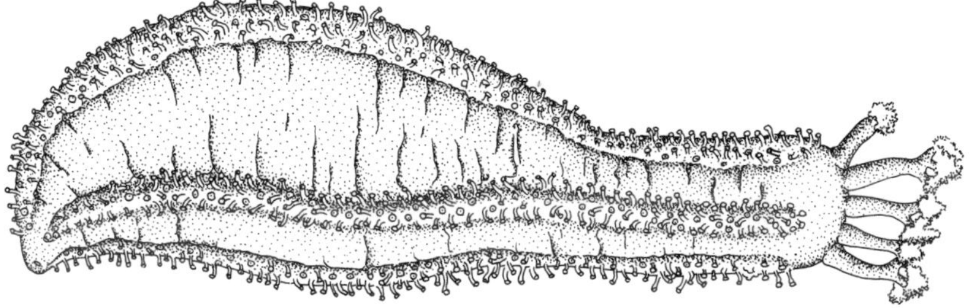

138. *Cucumaria miniata* (~ 15cm long)

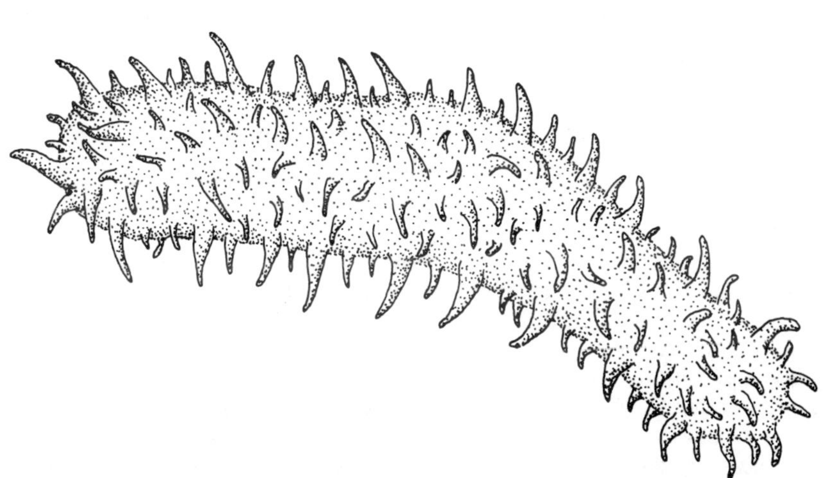

139. *Parastichopus californicus* (~ 20cm long)

Cucumaria lubrica is a small cucumber, generally less than 3cm in length. Local specimens are charcoal grey to nearly black, with a pale yellowish underside. Look for this species among mussels, or attached to lower intertidal rocks. Similar cucumbers, but without the yellowish color, are probably *C. curata*.

The largest of all holothurians is the giant *Parastichopus californicus* (fig. 139), reaching lengths as great as 50cm. These animals lack tube feet on their upper sides, but the mottled reddish body is beset with numerous fleshy spike-shaped projections. *Parastichopus* is typically found in subtidal habitats, but may occasionally be seen intertidally. When attacked by a predator, such as certain seastars, *Parastichopus* will often utilize its powerful body muscles to twist back and forth rapidly and engage in a primitive sort of swimming as an escape mechanism.

PHYLUM CHORDATA (with a notochord)

The chordates include the very familiar members of the subphylum Vertebrata (fishes, amphibians, reptiles, birds, and mammals), most of which are united by the possession of a backbone composed of vertebrae. The backbone, however, is not a requirement for inclusion in the phylum, and there are many invertebrate chordates. Membership in the phylum is granted those animals possessing a rod-like stiffening structure running most of the length of the body called the notochord (which is replaced during embryogeny by the backbone in many vertebrates), gill slits or holes in the walls of the pharynx or throat region of the digestive tract, and a hollow nerve cord placed dorsally along the body. These structures need not be present in the adult, nor all at the same time, for an animal to qualify as a chordate.

Subphylum Urochordata (=Tunicata)
the sea squirts and their relatives

There are three groups of urochordates in the sea, but only one is found intertidally. The classes Thaliacea and Larvacea are fully planktonic groups. The former includes jelly-like animals called salps, which float about singly or in chains and filter food from the waters of the open sea. The latter group is composed of odd little creatures which appear never to develop beyond the larval stage in structure, hence the name larvaceans. Of concern to us here is the Class Ascidiacea, the ascidians or sea squirts, all of which are sessile. These animals attach to a variety of substrates growing singly (solitary ascidians, fig. 140), in clumps of several individuals attached to a common base (social ascidians, fig. 141), or as masses with hundreds of minute individuals imbedded in a common jelly-like matrix (compound ascidians). Figure

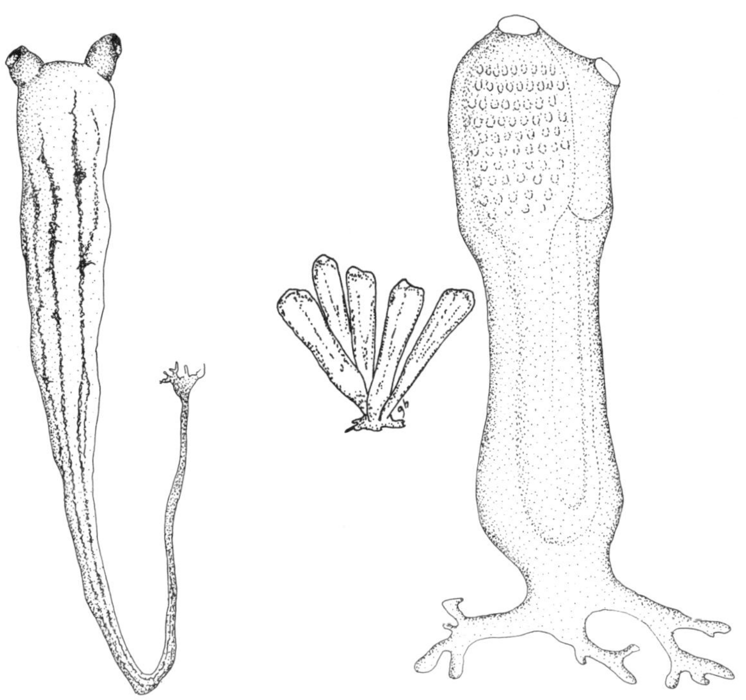

140. *Styela montereyensis* (~7cm long)

141. *Clavelina huntsmani* (zooid ~2.5cm long)

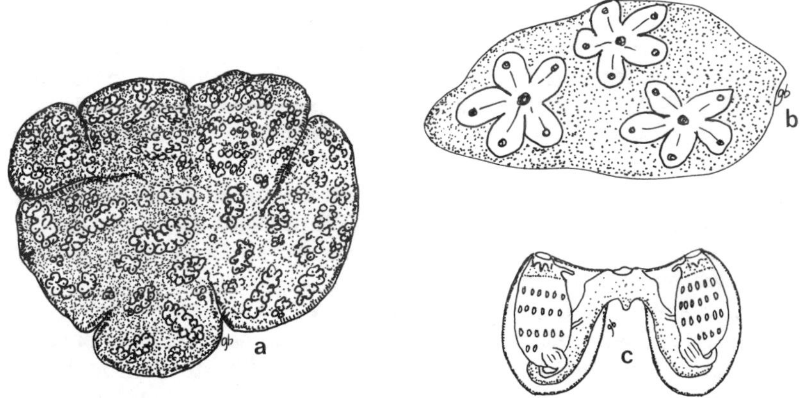

142. a-b, two compound ascidians; c, two zooids sharing a common excurrent pore (greatly magnified).

142 illustrates two zooids removed from such a compound type of ascidian. Each individual ascidian possesses an incurrent and an excurrent opening (siphons) through which water enters and leaves the body. In some species several zooids may share a single excurrent opening (fig. 142). As the water moves through the body it passes through a porous filtering basket formed from the pharynx in the front part of the digestive tract. The water moves through the pores to be carried out of the body via the excurrent opening, while food particles are trapped inside the pharynx in a sheet of mucus. The mucus and captured food material are then moved down the gut tube for digestion.

The body covering is called the **tunic** (hence the name tunicate) and may be slick and smooth, or rough and wrinkled. One of the main constituents of the tunic is a form of cellulose, a material characteristic of plants.

The most common solitary ascidian on the north coast is *Styela montereyensis* (fig. 140). These animals are quite large, ten or more centimeters long, and may be seen attached to the sides of large rocks, or hanging by their narrow stalk from the undersides of ledges in the low-intertidal zones. They prefer protected areas, and you are also sure to find some attached to floats in Humboldt Bay. The brown to reddish tunic is thick, leathery, and wrinkled and may sport a variety of small algae or encrusting animals growing on its surface. Close inspection will reveal the two siphons at the upper end, which are closed when the animal is disturbed or not under water.

A non-stalked solitary ascidian called *Pyura haustor* may also be seen among the tangles of algal holdfasts or other such protected areas. This animal is generally so overgrown with small plants and animals that the only visible parts of the body are the two bright red siphons which retract rapidly when touched.

Clavelina huntsmani (fig. 141) is probably the only clump-form or social ascidian you will see along our coast. Clusters of a dozen or so individuals are not uncommon in the well-washed channels of the lower intertidal zones. The tunic of these beautiful animals is clear and glass-like, while the internal pharynx bears a bright pink line, giving these animals a distinct and unmistakable appearance. The use of a hand lens when viewing *Clavelina* will help you to see some of its innards through the transparent covering.

By far the most abundant and diverse ascidians are the compound types (fig. 142). They are usually quite difficult to identify to species, and most people leave that task to the experts. Compound ascidians are often mistaken for sponges or thick bryozoan colonies. Look closely, however, and you will see the tiny individuals (**zooids**) imbedded within the jelly matrix. These ascidians do not feel like sponges or bryozoans either, for they tend to be slick or slimy to the touch. Most of these compound ascidians form rather thick masses usually conforming to the general shape of the substrate to which they are attached. In addition, they are often brightly colored, commonly orange or yellow, making them easy to spot among the low intertidal rocks, or on floats in bays. Some particularly beautiful types occur on pilings, floats, and eelgrass in Humboldt Bay. These are members of the genus *Botryllus,* which are dark purple to blackish with the tips of the zooids golden, or which are yellow, orange, or nearly red. The zooids of *Botryllus* are arranged in a regular star-shaped pattern around a common excurrent siphon. Another form, particularly common in Humboldt Bay, is *Botrylloides,* which is usually yellow or orange, with the zooids arranged in long parallel rows, with common openings occurring at intervals between them. The taxonomy of these beautiful and distinctive ascidians has not been entirely worked out, but it seems that several species of each genus may exist locally.

CHAPTER VII

THE VERTEBRATES

The three main groups of vertebrates associated with local shores are, of course, the fishes, birds, and mammals. We quickly omit the birds from our coverage, for they occur in such numbers and variety along the shore that to do them justice would require far more space than is available here. In addition, there are many other books devoted strictly to birds; we suggest Roger Tory Peterson's **A Field Guide to Western Birds**, as an excellent starting place for identifying shore birds. Included below are some of the more common fishes of our shores, and most of the marine mammals likely to be seen locally.

FISHES

Fishes are defined as aquatic vertebrates which utilize gills to exchange gases with their environment, and possess both paired and unpaired appendages (fins). Incidentally, the word "fish" is both singular and plural when referring to more than one fish of the same species; "fishes" is plural when referring to more than one species. Fishes are coelomate creatures, as are all of the vertebrates, and they possess well-developed organ systems for their various body functions. Three classes of living fishes are represented locally. Among the most primitive fishes are the lampreys (Class Agnatha, without jaws). The Pacific Lamprey, *Lampetra tridentata*, is common all along the California coast, and reaches lengths of over .5 meter. These odd, eel-like creatures have a circular, sucker-like mouth equipped with sharp hooks used to attach to the surface of other fishes. Once attached, the lamprey rasps a hole in the victim and feeds on the soft tissues and body fluids. Many salmon taken in local waters bear the scars of encounters with lampreys.

The Class Chondrichthyes includes those fishes whose skeleton is composed of cartilage rather than hard bone; the name means "cartilaginous fishes." The most familiar members of this group are the sharks and rays. Humboldt Bay abounds with a variety of bottom-feeding sharks and rays, and you may catch a glimpse of one as it cruises the bay channels during low tide. Among the loveliest of the local sharks is the leopard shark, *Triakis semifasciata*, which is tan with large dark spots, and may reach a length of nearly two meters. These animals, and most other bay sharks, cruise over or rest upon the bottom sediments and feed on a variety of invertebrates and small fishes. Most sharks brood their young in the body of the female, each fetus being provided with a yolk sac. Thus sharks give birth to their young rather than laying eggs as do most other fishes.

The large and beautiful bat ray, *Myliobatis californica*, is another common member of the Chondrichthyes in Humboldt Bay. These impressive animals often exceed a meter in width and weigh as much as 90 kilograms, but smaller individuals are much more common. The "wings" of this and other rays are actually modified pectoral fins which are used in swimming and also to stir up the bottom sediments and dislodge potential prey. The jaws are modified for crushing, and they are strong enough to mash even heavy-shelled clams. However, the northern author analyzed the stomach contents of nearly one hundred bat rays and found that they had fed largely on soft-bodied animals, mostly worms. These rays had been killed by people who feared that the big fish were eating oysters. Bat rays have a large spine on the dorsal surface of the base of the whip-like tail, which can inflict a nasty and easily infected wound.

The Class Osteichthyes (bony fishes) includes those fishes which have a skeleton composed of true bone. Humboldt Bay and the nearshore waters off northern California support a fantastic variety of these fishes, which in turn supports large commercial and sport fisheries. Those of greatest commercial and sport value are the king salmon (*Oncorhynchus tschawytscha*), the silver salmon (*O. kisutch*), various species of rockfish (*Sebastes* and *Sebastodes*), and numerous flatfishes such as flounders. Our concern here, however, is not with catching, killing, and eating these fishes, but with the many truly beautiful and fascinating fishes which can be seen and studied in the intertidal zone. A few of the more common types are discussed and figured below.

Several species of the Family Cottidae (sculpins) may be found in rocky intertidal pools along our coast. Most of these are commonly called "tide-pool johnnies" (fig. 143) and are in the genera *Oligocottus* and *Clinocottus*. These fishes are usually less than 10cm long and display a wide range of color patterns from grassy green to mottled red and brown. They are usually found to match closely the background on which they live. They also avoid predation by behavioral means; they may move quickly across a pool and then stop so suddenly that one's eye keeps going and misses these very hard-to-see animals. If you sit quietly beside a tidal pool, you are sure to see some of these little fishes darting about in search of food. Drop a small wriggling worm into a pool and a "johnnie" may sprint out for the morsel then back into some place of hiding. Predation by ambush is their typical style of feeding, although they will also sneak up on prey and snap it up very rapidly. They are known to prey on small crustaceans, worms, and snails, and also to graze on algae.

The staghorn sculpin (*Leptocottus armatus*, fig. 144) is common in Humboldt Bay. Most individuals are from 15 to 20 cm in length, grey on the dorsal side and yellowish-white below. Their common name is derived from the presence of sharp spikes which may be erected just in front of the gill openings. Although crustaceans are their preferred food, these carnivores will eat nearly any available animal material of appropriate size.

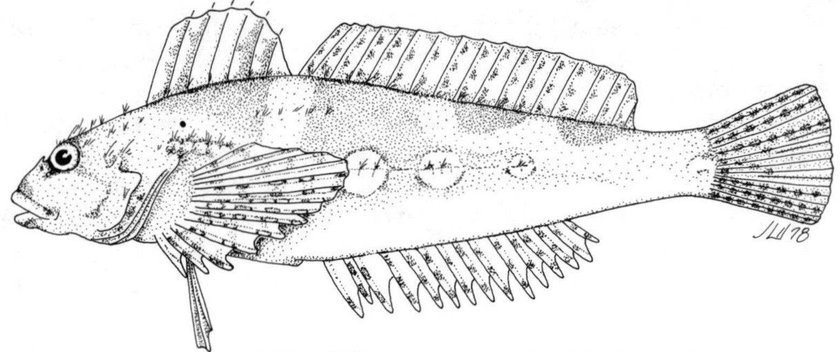
143. *Oligocottus snyderi* (~15cm long)

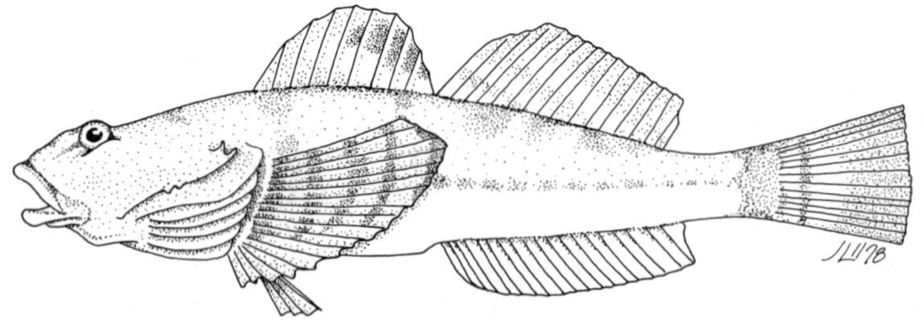
144. *Leptocottus armatus* (~15cm long)

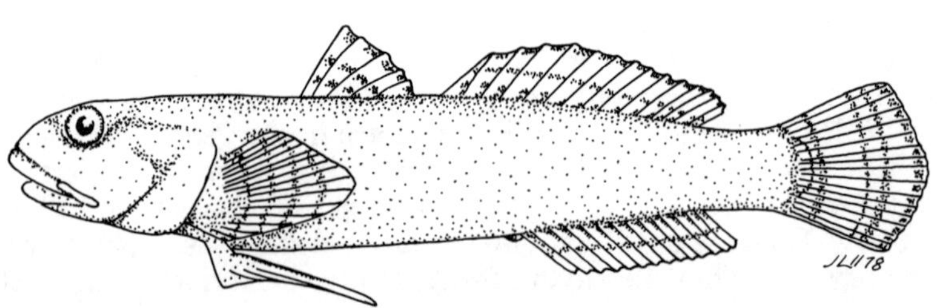
145. *Clevelandia ios* (~ 4cm long)

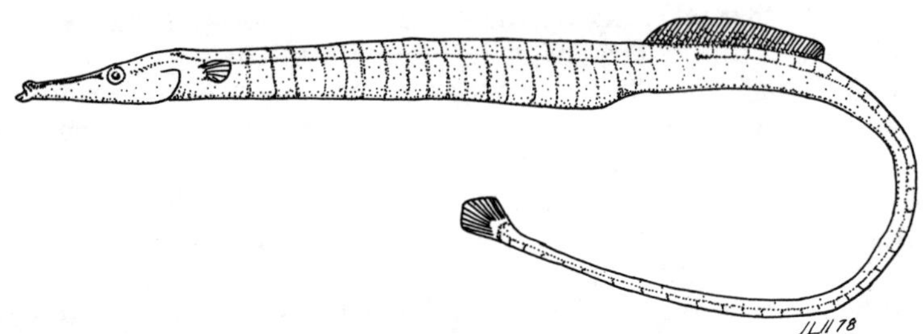
146. *Syngnathus leptorhynchus* (~18cm long)

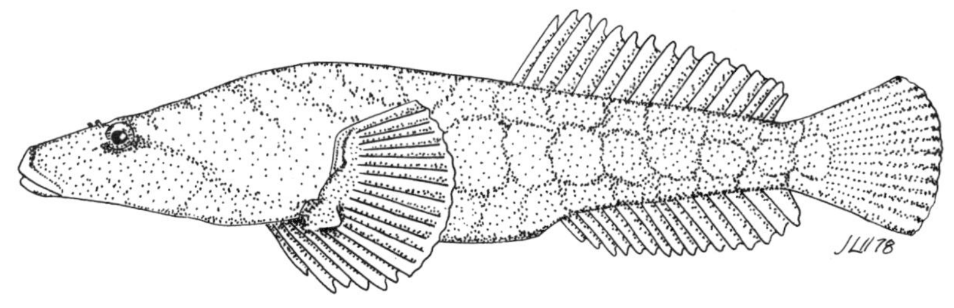

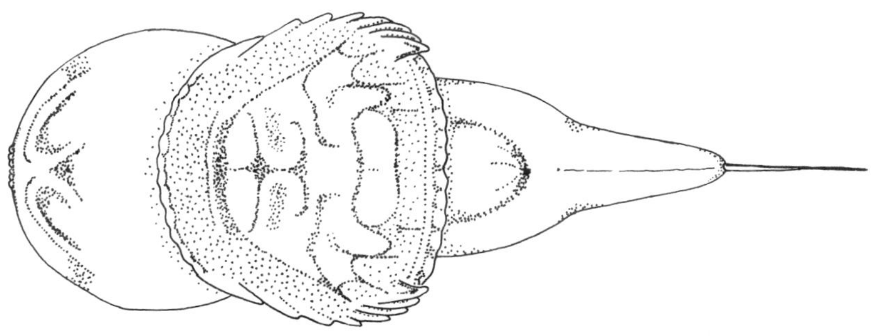

147. *Gobiesox maeandricus* (~5cm long)

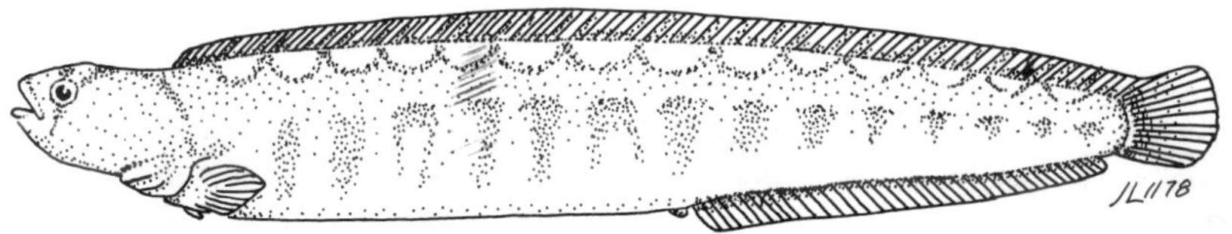

148. *Pholis ornatus* (~15cm long)

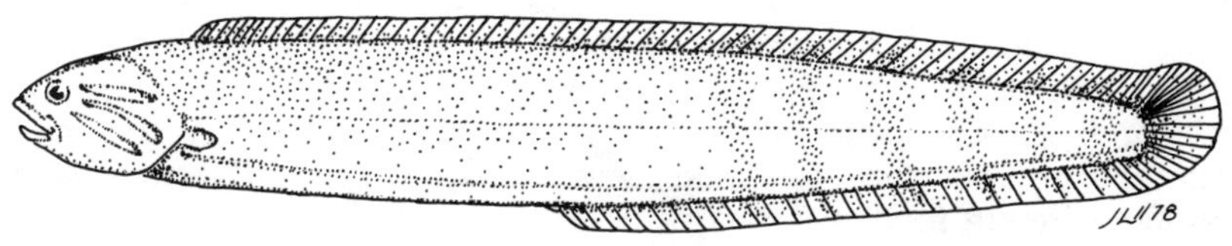

149. *Xiphister mucosus* (~15cm long)

Among the many other common fishes in Humboldt Bay is the arrow goby, *Clevelandia ios* (fig. 145). This species has been mentioned earlier as a resident in the burrows of *Urechis caupo*, the "fat innkeeper" worm. *Clevelandia* (Family Gobiidae) is generally about 3-5 cm long and may be found by digging in bay muds. As mentioned earlier, this fish usually leaves the burrow to forage for food during high tides, and then returns to its protected home as the water recedes.

Within the channels and eelgrass beds of Humboldt Bay you may find one of the strangest fishes of our coastal environment. The bay pipefish, *Syngnathus leptorhynchus* (fig. 146), belongs to the family Syngnathidae, which also includes the familiar seahorses. Bay pipefish are generally about 15-20cm long and vary from dark brown to grassy green in color, strongly resembling eelgrass. The resemblance is increased by the behavior of the fish, which orient their bodies parallel to the blades of the eelgrass and wave back and forth in the water with the movement of the leaves. The favorite food of this fish is small crustaceans. The tiny mouth at the end of the elongate "snout" can suck tiny crustaceans in from as far away as two or three centimeters so rapidly that one cannot follow the action. As in the seahorse, it is the male pipefish which carries the developing young in a brood pouch or **marsupium.**

As you turn over stones in rocky intertidal areas, you are likely to find a small, somewhat tadpole-shaped fish attached to the rock surface by a sucker on its belly. This fish is *Gobiesox maeandricus*, the northern clingfish (fig. 147), in the Family Gobiesocidae. The sucker is formed from highly modified, fused pelvic and pectoral fins (the paired fins). These 5-10cm-long fish may survive several hours of low-tide exposure to air by remaining in damp areas under rocks. During high tides they swim about the protected pools where they feed primarily on small crustaceans.

Look under rocks and in tidal pools for eel-like fishes commonly referred to as "blennies" (several groups are included undered this vernacular name; not all blennies belong to the family Blenniidae). Most of them range between 10 and 20cm in length. The two most commonly represented families on our coast are the gunnels (family Pholididae) and the pricklebacks (family Stichaeidae). Two local examples are shown in Figures 148 and 149. The two families may be distinguished from one another by the relative lengths of the anal or ventral fins (the elongate, unpaired fin on the ventral side of the body) as illustrated. These fishes are also common in eelgrass beds of Humboldt Bay. Like so many intertidal fishes, the gunnels and pricklebacks are capable of surviving periods of exposure during low tides by remaining in protected moist areas. Also, many of them secrete mucus over their skin to reduce desiccation further. The gunnels and pricklebacks feed on a variety of invertebrates, including crustaceans and small mollusks, and also on algae, different species having different food preferences.

150. *Eumetopias jubata*, Steller's Sea Lion, male.

MARINE MAMMALS

The most often-seen mammals of our north coast are the playful seals and sea lions. Both of these types of mammals belong to the Pinnipedia. The name is Latin for "feather feet," in obvious reference to the limbs, which are modified as flippers. In addition, the pinnipeds have a very short, vestigial tail and sleek streamlined bodies that make them hydrodynamically adapted for their aquatic life styles. The skin of seals and sea lions is quite thick, generally well haired, and underlain by an even thicker layer of fat (blubber), all providing insulation that allows these animals to retain their body heat in the cool waters which serve as their home. But good insulation can cause problems if the animals become overheated (as from exertion), and they need a mechanism to rid the body of excess heat. To overcome this problem they have certain regions of the skin, particularly on the flippers, provided with well-developed networks of blood vessels.

When the body overheats, blood passing rapidly through this system near the body surface is cooled. Blood flow to these special areas is decreased or increased, depending on whether the animal needs to utilize its "heating" or "cooling" systems respectively.

The pinnipeds are a fairly young group, as evolutionary time is measured, that probably evolved about 40-50 million years ago. They are closely related to the Order Carnivora (the bears, sea otters, dogs, etc.). Of the three families of pinnipeds, the walruses (Odobenidae) and sea lions (Otariidae) probably evolved from a bear-like ancestor, and the seals (Phocidae) from an otter-like ancestor. Together these groups consist of only about 30 species, six of which occur in California, and three of which frequent the north coast.

The sea lions (and fur "seals") are recognized by the presence of small (but distinct) external ear flaps and the ability to fold the hind flippers forwards, alongside the body. These are the agile creatures seen in various trained "seal" acts, and most species are known to be quite intelligent and communicative. They all possess a well-developed social system that is most evident when they come ashore for their breeding periods. Two species of sea lions may be seen along our north coast, Steller's sea lion (*Eumetopias jubata*, fig. 150) and the California sea lion (*Zalophus californianus*, fig. 151).

151. *Zalophus californianus*, California Sea Lion, male and female.

Steller's sea lions (also known as northern sea lions) are large animals, with males reaching nearly 4m in length and weighing to 900 kg (2,000 pounds), and with females reaching over 2.5m in length and weighing about 460kgs (1,000 pounds). They range throughout the shores of the north Pacific from the southern Channel Islands to the Bering Sea, and along the Asian coast to Japan. These spartan animals show a particular preference for the rugged shores and offshore rocks of the outer coast which makes them difficult to locate and observe. In our area they can best be viewed along Abalone Beach south of Patrick's Point, or on the offshore rocks near Trinidad. In early summer the large bulls begin "hauling out" onto deserted beaches to establish their breeding territories and attract their harems, which will eventually contain from a few to as many as 30 cows. The territorial battles between competing bulls usually get quite vicious and noisy with their constant barking. Experienced males often bear bloody bites and wounds; the battle scars of their instinctive drive to obtain beach space. These handsome animals are known to consume large quantities of fishes, including many so-called "trash fish" (those of no commercial value), as well as salmon, and lampreys associated with the salmon. They may dive to depths of over 165m (about 500 ft) during their feeding forays.

California sea lions are darker in color than the Steller's and noticeably smaller, males reaching about 2.5m in length and weighing about 265kgs (600 pounds), females being usually less than 2m long and weighing about 90kgs (200 pounds). Only the males occur further north than about San Francisco Bay. These animals are distributed in three geographically isolated populations, each being designated as a separate race or subspecies. One race lives on the Hunshu Islands of Japan and is precariously close to extinction, with a total population of only 200-500 individuals. A second race exists only in the Galapagos Islands and is made up of about 20,000 animals. The third race ranges from northern California south to the Gulf of California, Mexico, and probably includes over 50,000 animals. Although this form (*Zalophus californianus californianus*) has been recorded as far north as Puget Sound, it is common only in central and northern California. On the north coast the males may be observed on the offshore rocks near Trinidad Head. Like the Steller's sea lions, the California sea lions usually avoid shores frequented by people. In response to human disturbance these creatures will often abandon their habitat for considerable periods of time, sometimes several weeks. If repeated disturbances occur they may abandon an area permanently. For this reason observations should always be made from a distance, with a spotting scope or binoculars, and one should try to be as inconspicuous as possible.

In order to assure contact between sexes in a population that is scattered over many hundreds of square miles, **rookeries** have evolved among the California sea lions. These are special breeding grounds mostly on the southern offshore islands, where these sea lions congregate in large numbers during the appropriate times of the year.

Breeding males, about five years of age and older, arrive first at the breeding grounds and battle among themselves until relatively few remain. The occupants, dominant bulls, do not hold a discrete area, but shift territories with time of day, changes in temperature, tidal level, and location of the females. Other fully grown, but subordinate, males may be tolerated within the territory of another male for several hours. The harem size of these California sea lions varies from 10-20 cows. Females begin to breed at about three years of age. Shortly after arriving at the rookery a mature female usually gives birth to a single pup. A few days to a week after the pup is born the cow comes into heat and mates. The gestation period is between 342 and 365 days and includes two distinct phases. During the initial phase of development, the embryo remains free in the uterus of the mother and is more or less dormant. Later, the second phase is initiated by implantation of the embryo in the uterine wall, following which development continues. Thus the "expected" gestation period is somewhat less than one year, but because of the delay caused by the first phase of "dormancy" birth does not occur until the onset of the next breeding time, synchronizing the two events. Pups are nursed for three to five months on a milk composed of about 15% fat and 19% protein, one of the richest among mammals. Because the milk is so rich the mother need not nurse the young very frequently, allowing her to spend more of her time in search of food (which becomes scarce where the sea lions are concentrated during breeding). The diet of the California sea lion is not unlike that of the Steller's; fishes, squids, and crustaceans form the staple meals. After the breeding season is over the sea lions scatter up and down the coast, and many males may subsequently be seen along our northern beaches.

The "true seals" (family Phocidae) are without external ear flaps and are hence more appropriately referred to as the earless seals (or better yet, the phocids). Unlike the eared seals (sea lions), the phocids cannot fold their hind flippers forward alongside the body. Where the sea lions take the weight of their bodies on the flippers when moving on land (and are quite agile), the phocids move in gigantic inch-worm fashion by taking their weight on the sternum (chest) and then the pelvis (hips). A beach-traveling phocid is a rather ungainly animal, but once in the water they become among the most graceful of all swimming creatures. There is only one kind of earless seal along the northern California coast, the handsome harbor seal.

Harbor seals (*Phoca vitulina,* fig. 152) are very sleek, cigar-shaped animals that frequent bays, harbors, river mouths, and local offshore rocks. Adult males may reach nearly 2m in length and weigh from 100-140kgs (about 220-300 pounds), females are slightly smaller. This species ranges on our coast from the Arctic Sea and the Aleutian Islands, south to central west Baja California. *Phoca vitulina* is a cosmopolitan species and also occurs in Europe, Asia, and on the eastern seaboard of the United States. The entire northeastern Pacific population of harbor seals has been estimated at about 100,000 to 200,000 individuals. Unlike the sea lions of our coast, harbor seals do not form

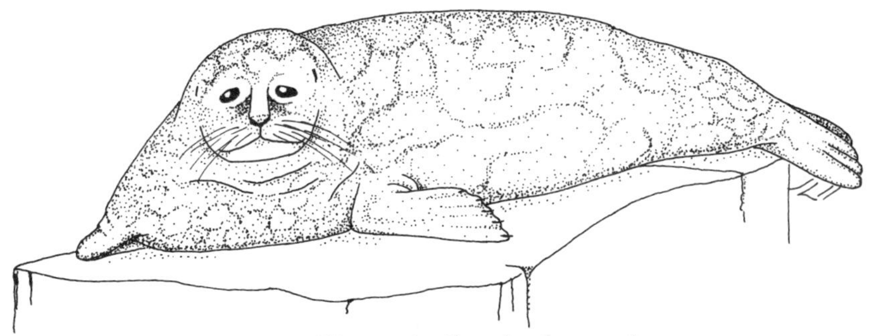

152. *Phoca vitulina*, harbor seal.

organized breeding harems, and both sexes appear to be quite promiscuous. Pups are born (on our coast) in the spring, and birth may take place either on the beach or in the water. The mother often leaves her pup on a secluded sandy beach or among protected rocky pools while she forages for food which consists largely of slow-moving fishes and benthic shellfishes. If you happen on one of these pups, realize that it is probably not abandoned for long and should not be touched or disturbed. On several occasions, well-meaning beachcombers have brought such young seals to the Marine Laboratory at Trinidad thinking the animal needed care. Since people cannot provide the care of the mother, the babies usually die. There is also some question as to whether a mother seal will accept her pup after it has been touched by a human, so please enjoy them from a respectable distance and let Nature take her course.

Harbor seals are often seen floating in the water with only their heads in view, sinking down out of sight just as the observer gets his or her binoculars focused. Although these animals are shy and difficult to approach, they are quite curious, and we've seen them follow us, offshore 30 or 40 meters, as we strolled down some California beach. Near Trinidad, Patrick's Point, Humboldt Bay, and in the mouths of local rivers they may be approached close enough for snapshots if one crawls along the low rocks or sand flats or quietly rows one's skiff.

There are two kinds of otters that occur on the California coast, the sea otter and the river otter. Both are members of the Order Carnivora and the family Mustelidae. The mustelids are a rather diverse group of mammals that also include the weasels, mink, wolverine, skunks, and a few others. All of these carnivores have fur of extremely fine quality, well-developed anal scent glands, and short muzzles.

River otters (*Lutra canadensis*) are long, elegant, graceful animals that occasionally swim out of the mouths of their home rivers to frequent local estuarine and bay habitats. Patient observers will eventually see these playful creatures swimming about in Humboldt Bay and near the Trinidad pier. Their silky bodies are about one meter long, with a thick tapering tail, long round head with prominent whiskers, and short legs with fully webbed feet. When in their freshwater environment, river otters feed upon "non-game" fishes, frogs, turtles, crayfish and insects; when in the sea their diet probably consists of slow-moving fishes and crabs, and they scavenge on dead fishes and fish parts (as around Trinidad pier). These are extremely intelligent creatures, and even a brief period of observing their behavior will thrill and excite the part-time as well as the professional naturalist. No doubt a great deal of the pleasure derived from watching river otters stems from the strong resemblance of their play activity to that of ourselves — particularly when we were children.

Sea otters (*Enhydra lutris*) are larger than their freshwater cousins, reaching over 1.5m in length, with their broad tail accounting for about one-fifth of that length. The fur of these animals is of so fine a quality it exceeds even that of the mink. Indeed, it was this attribute that brought the sea otter nearly to extinction just a few decades ago. During the latter part of the 19th century they were hunted relentlessly, their population diminishing yearly. It has been said that their near extinction in Alaska had substantial influence on the Russian decision to sell that land to the United States in 1867. It was not until 1911 that an international agreement was signed to protect the sea otter, and by that time its once extensive range had been reduced to a small herd along the central California coast, and perhaps a few individuals in the most remote of the Aleutian islands. In the years since their protection they have extended their range only slightly. They may be seen with increasing frequency in Monterey County, but if you spot one along our coast the sighting should be reported to the Department of Fish and Game. Thus, while sea otters do not presently occur (or at least they are not seen) along the extreme north coast, they may have occurred here originally.

Sea otters are largely offshore creatures and are seldom seen hauling out on beaches or even isolated offshore rocks. They swim, or float on their backs, often living in close proximity to kelp beds. It is not uncommon, in fact, for a sea otter to tie itself up with kelp while it sleeps, presumably to avoid drifting away or into the surf. The otters feed extensively on shellfishes, including sea urchins, clams, and abalones. The somewhat hysterical claims of commercial abalone fishermen appear to be without foundation, for it is known that abalones were quite abundant during earlier times, when the size of the sea-otter population was far in excess of what it is now. The reason there are so few abalones left in California is simply because people have over-harvested them. In addition to this, it has been shown that where sea otters do occur, their predatory patterns are similar to those discussed in the ecology sections of this book as "indiscriminant top carnivores" — that is, their presence in an area actually increases the diversity, health, and biological complexity of shallow-water marine communities. They may also be important in maintaining natural kelp beds, by virtue of their predation on kelp-destroying sea urchins (see Chapter III). Sea otters are one of the few mammals, other than the primates (which include man), which use tools. To gain entry to certain hard shellfishes the otter will crack them open with rocks held on its chest, while floating on its back. The primary enemies of these unique animals, the only living members of the Order Carnivora to have evolved an entirely marine life style, are killer whales, some sharks, and of course man.

The whales, porpoises, and dolphins constitute the Order Cetacea. The cetaceans are extremely specialized aquatic mammals that have lost the typical body hair and associated epidermal glands of all other mammals, eliminated all external vestiges of the hind limbs, and modified the forelimbs into paddle-like flippers. In addition, most have evolved a dorsal fin, like that of fishes, and the nostrils have moved to the top of the head where they form the "blowhole." The major evolutionary steps that led to these changes probably occurred a little more than 70 million years ago. Like the seals, the cetaceans have a layer of insulating blubber beneath their skin, but since they lack the heat-retaining fur of the pinnipeds this blubber layer has become impressively well developed, in some species accounting for more than one-third of the animal's total weight. Other specializations for their aquatic existence have taken place in the lungs and circulatory system. These modifications allow some large whales, such as the sperm whales, to remain submerged for more than an hour, while they dive to depths of nearly 1,000 meters (over 3,000 feet!), where the water is perpetually cold and black. Swimming is accomplished by vertical movements of the tail fin or flukes, while the flippers and dorsal fin act as stabilizers.

If "relative brain size" (i. e. brain mass per body weight) is used as a measure of intelligence — and increasing amounts of evidence suggest that such a relationship does indeed exist — man is seen to sit atop the scales. But immediately below man, and above the non-human primates (chimpanzees, baboons, etc.) are the dolphins. We doubt that there are many students of mammalian intelligence who would deny that man and dolphin are among the most intelligent life forms on Earth today. A point of interest is that, among the fishes, the predatory sharks appear to have the greatest relative brain sizes. It has been suggested by some that porpoises are nearly as smart as people, but we may never discover just how intelligent they are because our communication systems are mechanically and conceptually vastly different. Almost certainly, humans and cetaceans are intelligent in different ways, and one can only speculate as to the selective forces that produced such a nervous system in the sea.

There are two groups of cetaceans: the baleen whales (suborder Mysticeti) and the toothed whales, dolphins, and porpoises (suborder Odontoceti). The baleen whales lack teeth and bear instead large plates which hang from the roof of the mouth and function as giant sieves or filters. These plates are called **baleen** or **whalebone** and are formed from modified layers of skin, something like fingernails. Whalebone is quite durable, yet flexible, and not unlike a frayed palm frond in appearance; it is the frayed edges that serve as the filters. There may be several hundred such plates in a whale's mouth. During feeding water is taken into the mouth and then forced out through the baleen by the tongue, and in the process small planktonic animals are captured in the filters. The size range of trapped animals is such that most food consumed consists of small crustaceans called euphausiids or **krill**, mysids, amphipods, shrimp, and small fishes. The baleen whales are further characterized by the retention of both nostrils and hence have what is referred to as a double blowhole. (Whales, by the way, do not inhale or exhale water, they are air-breathing mammals. The water seen to "spout" from the blowhole as one of these leviathans surfaces is merely a stream of vapor, arising from water trapped around the outside of the nostrils and, in colder weather, condensation of water from the whale's moist and lung-warmed air as it hits the colder atmosphere, just as our own breath fogs on cold mornings). The Mysticeti are large whales and include the largest animal ever to live, the giant blue whale *(Balaenoptera musculus)*, a cosmopolitan form reaching lengths of about 30m (100 ft) and weights of nearly 150 tons. Of the 10 living species of baleen whales only one is commonly seen along our north coast, the California grey whale.

Viewing the California grey whale (*Eschrichtus robustus*, fig. 153) has become somewhat of a tradition in California, because of its annual migration along the coast. The grey whale leaves its summer feeding grounds around the Aleutian Islands and Chukchi Sea to journey 5,000 to 6,000 miles south to its ancestral breeding grounds along the shores of Baja California. This incredible journey takes about three or four months in either direction and covers 50 degrees of latitude; it is the longest known migration by a mammal. The first migrants to abandon the Arctic feeding grounds each year are the pregnant females, followed shortly by the males, young, and non-pregnant females. These animals leave the Arctic around September, moving southward. They follow the coastline closely, oftentimes being within a half mile of the shore, apparently utilizing local topography and currents to navigate. Islands and large coastal points and offshore rocks appear to be major reference points during the journey. They average 4 to 5 miles per hour as they swim, covering about 100 miles each day. They probably swim both day and night, communicating all the way by the mysterious grumbles and whistles that are the language of whales. By late December and early January most of the whales, in groups of one to six animals, have arrived at the breeding grounds, although some may be seen through February off northern California. Some of these stragglers may in fact never make it all the way to Mexican waters. The principal breeding and calving areas are the warm shallow lagoons of western Baja California (Laguna Guerrero Negro, Scammon's Lagoon, Bahía San Ignacio, and Bahía Magdalena), but a few persistent individuals push further south to the lower Gulf of California, utilizing Laguna Yavaros and Bahía Reforma. These lagoons and bays are all on pristine and isolated desert shores, and it is for this reason that the grey whale is sometimes called the "desert whale". This species has also been referred to as the "hardhead" and "devilfish". The entire journey takes these animals through environments ranging from 0 °C (with 22 hours of daylight) to 18 °-22 °C (with 8-9 hours of daylight).

153. *Eschrichtus robustus*, California grey whale.

Both breeding and calving take place in the Mexican lagoons, and the newborn calves (and their parents) remain there until March or April, when they begin the long trek northward. The newly pregnant females leave early, during February and March, for it is important that they have the longest possible feeding period to support the "two of them". Unlike other baleen whales, the grey whale appears to be exclusively a bottom feeder, mucking about in soft silty and muddy sediments to filter out the benthic animals which live there (primarily small crustaceans).

There are actually two separate groups of grey whales, the California population and the Korean population. The latter is a small cluster that represents a mirror image of our own population, migrating south to breed in the warm waters between Japan and the Asian mainland each winter, and north to subarctic water to feed during the summer.

Since these beautiful whales were placed on the protected species list the only predator of the grey is the killer whale. Numerous scars on the flukes, flippers, and elsewhere suggest that predatory attacks are common but serve primarily to remove only the old or sick individuals, and possibly some young.

For a number of years the southern author worked as a naturalist and guide for groups traveling to Baja California to study the grey whale (and other species occurring in western Mexico) during their breeding and calving periods. During these trips he had occassion to encounter an intrepid Baja traveler by the name of Margie Stinsen, who told him an interesting folk tale concerning this cetacean's migratory instincts. It would seem that the whales' drive to return to the place of their birth is indeed a powerful one, and upon death even their spirits desire to return to rest on the Vizcaino Plain of Baja California. If they die far from this land, they must float beneath the waves and earth back to the lagoons, and on windy days one can see the spirits of many whales blowing spouts of dust devils as they try to traverse the land on their long journey home.

On our north coast the best times to watch the migrating grey whales are from April through June, and from early December through early February. Any high cliff area which juts out into the sea will serve as a vantage point for whale watching. Wedding Rock at Patrick's Point and the cliffs around Trinidad are fine local viewing spots. The Humboldt State University Marine Laboratory at Trinidad keeps records of sightings, and the staff can tell you where recent observations have been made.

The toothed whales (Odontoceti), in contrast to the baleen whales, have teeth and no baleen and the nostrils fused into a single blowhole. In addition, the toothed whales are generally smaller than the baleen whales and include the smallest whales although a considerable size range exists, from porpoises and dolphins, 1.5 to 2m long, to the giant sperm whales which may reach lengths of nearly 20m (60ft.). There are about 75 living species of toothed whales, 18 or 19 of which occur in California waters, two of which may be seen from shore in northern California: the killer whale and the harbor porpoise.

Killer whales (*Orcinus orca*, fig. 154) are fairly large cetaceans, males reaching nearly 10m (30ft) in length, and females 7 or 8 meters (23 ft.). They are commonly sighted in pods (packs) of from two to six individuals, although counts of 200 in a pod have been recorded in some areas. They are world-wide in distribution. Killer whales are quickly recognized by their prominent dorsal fin which reaches heights of 5 or 6 feet in males and usually sticks straight up. In females the dorsal fin is slightly shorter and curved posteriorly. The individual shown in Figure 154 is a female and a personal aquaintance of the artist, Ms. Jody Hawkins. Killer whales have a distinctive smooth white belly and a white patch behind their tall dorsal fin. these cetaceans are unquestionably one of the handsomest of all creatures in the sea, and their sleek, well-formed bodies move through the water

154. *Orcinus orca*, killer whale

with a power and grace nearly unmatched by any other animal on Earth. The tall dorsal fin cuts a smooth rapid slice through the sea's surface as these whales travel mile after mile on journeys still not understood by man. *Orcinus* is also one of the most intelligent life forms on our planet, and the small amount of reasearch that does exist, based primarily on captive individuals, suggest that these cetaceans have extremely sophisticated social and communicative behaviors. Field observations on killer whales indicate that they hunt in packs, comparable to those of wolves. They display a high degree of cooperation and almost certainly communicate during hunts, with some acting as scouts, others as herders, and others as killers. There is still, of course, a great deal of controversy over how complicated these behaviors really are, especially the aspects of communication and division of labor. Humans often seem to resent the idea of other creatures being intelligent and capable of anything approaching rational decision-making or thought.

Killer whales appear to prefer warm-blooded prey and frequent pinniped rookeries where the young seals and sea lions are easy prey. They are also known to attack baleen whales, walruses, penguins, and sea birds. Fishes, turtles, and squids are also included in their diet to lesser degrees. There are no authenticated reports of attacks on humans in the sea. Their powerful jaws are each supplied with 10-14 pairs of massive conical teeth. Despite their beauty and grace, intellect, and apparent "hands-off" policy with regard to people, they are indeed appropriately named — the rapidity, strength, and finesse with which they attack their prey almost always result in a successful kill. Local sightings of killer whales near shore are irregular and mostly by chance.

Members of the porpoise family (Phocoenidae) do not have the elongated snout characteristic of most marine dolphins (Delphinidae), nor are they as large. In addition, the phocoenids have flattened or spade-like teeth, as opposed to the conical teeth of most dolphins. Only a single porpoise (and no dolphins) is commonly seen near shore along the north coast, although other species may be encountered at sea. The harbor porpoise (*Phocoena phocoena*, fig. 155) is the smallest cetacean known from California waters, reaching about two meters (6 feet) in length and weighing about 70-75 kgs (160 pounds). Its dorsal fin, in contrast to that of the killer whale, is quite low and triangular, and the body lacks the distinctive markings of the killers, being black on top and shading to white below. Very little is known about this cetacean.

155. *Phocoena phocoena*, harbor porpoise.

It ranges from Alaska to central California along our coast and has a habit of frequenting bays, harbors, and other inshore waters. The patient observer has a good chance of seeing these animals in Humboldt Bay, especially from a vantage point overlooking some of the deeper channels. They generally swim in pairs or trios, with their black sleek bodies rolling at the water line as they cruise the bay channels. The small amount of work conducted on these animals indicates that they feed primarily on small fishes and squids.

Last, but not least, there is a clever little terrestrial mammal that frequents our shores, particularly during low tides at night — the raccoon, *Procyon lotor*. Raccoons are common throughout California and, of course, many other states, and are both resourceful and flexible in habit. They are known to feed on a great variety of meats, insects, fruits, berries, eggs, and human foodstuffs (from the backyard chicken coop or garbage can). They are extremely curious animals and have a habit of handling their food for a considerable length of time prior to eating it. It is often stated that they "wash" their food, but in fact it is probable that the water increases their touch sensitivity and that they are "feeling" their food during the handling process. It may be that they prefer particular textures or consistencies over others in their food selection. During their nighttime seashore forays their principal food from the marine menu is crab, but they also search through washed-up algae for other crustaceans. One can always tell when a raccoon has been dining at a local seashore, for the remains of crabs will be scattered about the rocks, and the morning sand will bear the predator's distinctive tracks, which like those of bears or people leave an entire impression from heel to toe. You may happen on a few of these animals among the rocks at early low tides — it is a good piece of advice to be on about your business and let them be on about theirs.

CHAPTER VIII

MARINE PLANTS

The most significant plants in the sea are not the obvious and abundant seaweeds and kelps seen along the shore, but rather the microscopic, single-celled forms, most of which are planktonic in offshore waters. One such group, the dinoflagellates (Division Pyrrophyta), has already been mentioned in connection with the occurence of "red tide". (The category of **Division** in botany generally corresponds to the level of Phylum as used in zoology.) A second group, and by far the more important, is the Division Chrysophyta, the diatoms. They occur in uncountable numbers (**bloom**) during optimal reproductive periods and are the main contributors to primary productivity in the sea's economy. They are eaten directly or indirectly by everything which feeds in the sea, and those land creatures which eat marine organisms. These little plants are encased in cell walls composed of silica (very similar in composition to glass); the ornate cases are formed as two halves of a tiny box which fit together around the cell. The "shells" of dead diatoms cover much of the ocean's bottom as an ooze. At Lompoc, California (near Santa Barbara), giant cliffs of these deposits, exposed over the centuries by coastline movement, are mined and the "diatomaceous earth" is used in a variety of industrial processes, especially in the manufacture of filters. Some diatoms are not planktonic, but rather are attached to substrates as a brown or yellowish film or scum, and are important as food for browsing herbivores such as limpets. Look on the blades of eelgrass, or on the south slope of ripple marks on tidal flats, for these films of diatoms. These creatures are extremely beautiful when viewed through a hand lens or low-powered microscope.

There are four divisions of conspicuous intertidal plants which you will encounter along our coast. These are the green algae (Chlorophyta), red algae (Rhodophyta), brown algae (Phaeophyta) and the eel- and surf-grasses (Trachaeophyta). The trachaeophytes include what are called the **vascular** plants, those with particular types of tissues for the conduction of food and water through the body, and which possess true roots, stems, and leaves. Many of the familiar land plants (trees and other flowering plants) belong to this division. The eel- and surf-grasses are specialized flowering plants (family Zosteraceae) which have become adapted to life in the sea. The eelgrass (*Zostera marina*) is restricted to soft-bottom, relatively stable habitats such as the tidal flats of Humboldt Bay. As noted throughout this handbook, eelgrass beds provide living quarters for a great variety of animals. The surf grasses (*Phyllospadix*) have narrower leaves than eelgrass and are commonly found along rocky shores. The pale yellowish-green flowers of both of these plants occur in double rows along thickened areas of the leaves. Pollen is released into the water and carried by waves and currents to other plants.

The algae are among the groups of non-vascular plants and lack true roots, stems, and leaves. Their bodies consist of some sort of **holdfast** which forms the attachment to the substrate, and a **thallus** or main portion of the plant. The thallus may consist of simple flattened **blades** resembling leaves, or it may have a stem-like **stipe** supporting the blades off the substrate. In some there is an obvious **midrib** down the middle of the blades. The thallus itself may be "leafy", encrusting, branched, or composed of fine filaments.

Although the diatoms are of greater importance in the overall economy of the sea, the dense growths of algae along the coast provide localized habitats, protection, and sources of food for many intertidal animals. These algae represent a tremendous amount of organic matter which contributes greatly to the energy budget of shore habitats. As has been pointed out earlier, a number of invertebrates and fishes are herbivores and graze on various species of algae. These animals, in turn, support the predators which are thus indirectly dependent on the plants. Many other animals feed almost exclusively on the organic scum which grows or accumulates on the blades of various intertidal plants. We have also seen that a host of invertebrates utilizes plants as homesites, living either in the tangled root systems of eelgrass and surf grass, or on the blades of algae, and many of these animals have evolved patterns of protective coloration in conjunction with these habitats.

The reproductive cycles of most algae involve two distinct phases: a sexual gamete-producing stage (called the **gametophyte generation**), and an asexual spore-producing stage (called the **sporophyte generation**). These stages typically alternate, one giving rise to the other. Generally speaking, the gametophyte stage produces gametes, fertilization takes place, and the resulting embryo develops into the sporophyte stage. The sporophyte produces spores each of which (without fertilization) develops into another gametophyte, completing the cycle. There is a tremendous amount of variation and complex "sub-cycles" within this basic plan, especially among the red algae, and some of the differences are important in the taxonomy of algae.

Throughout the evolution of the plant kingdom there has been a tendency towards the domination of the life cycle by the sporophyte generation in the higher forms. For example, the obvious flowering plants and cone-bearing trees on land are sporophytes, while the gametophyte generations of these plants are represented only in the pollen grains and seeds. The same is true for many algae; most of the obvious plants are sporophytes. Some exceptions exist, of course. For instance, the gametophyte and sporophyte generations of the green alga *Enteromorpha* are indistinguishable from one another without resorting to detailed microscopic examination.

Seaweed life cycles along our coast, and many other areas of the world, are drastically affected by seasonal changes. Growth and reproduction are dependent upon available nutrients and sunlight (day-length) which are necessary for photosynthesis, so that dense beds of algae are present in most areas during the summer, while some regions appear barren during the winter months. This seasonal difference is made even greater by physical factors such as heavy winter surf which abrades many algae from the rocks. Some of the algae (*Egregia*, *Alaria*, etc., figs. 163-164) partially compensate for adverse winter conditions by having holdfasts which are extremely tough and can persist even when the rest of the plant has been torn away. In the springtime these holdfasts begin to grow new blades and thus get a "head start" on those algae which must resettle in the area. Summertime growth rates of some algae (e. g. *Alaria*) are phenomenal, up to several inches each day!

Certain species of algae are of commercial importance in various parts of the world. *Porphyra*, for instance, is grown and harvested in the Orient as a food used in Japanese cooking. Closer to home, the giant kelp beds of southern California (which consist primarily of *Macrocystis pyrifera*) are harvested commercially for an extract generally referred to as **algin**. This material is used in the manufacture of a great variety of products. Certain chemical forms of algin provide a medium for the culture of bacteria, while others are used as emulsifiers to prevent separation in such things as ice cream, cosmetics, salad dressings, and the like. Read the labels on these products and you are certain to see "algin extract" or "sodium alginate" listed in the ingredients. For those of you interested in more information on these and other algae, we refer you to the reference list for appropriate titles.

Some algae, especially the "reds", are very difficult to identify, and we have included only a very few of the many species found along our shores. We again caution the reader to refer to some of the appropriate volumes listed in the references if a plant does not precisely fit our description or figure. It is not always easy to decide to which division a particular plant belongs, but the following rule of thumb, based upon color, will be of some help (the actual differences between divisions lie in the types of pigments, details of reproductive structures and life cycles, and the chemistry of storing food.) Assuming that the plant in hand is not one of the "grasses" mentioned above, if it is bright or mossy green in color it is almost certainly a member of the Chlorophyta. Most green algae are relatively small, with thalli rarely exceeding one foot in length. Any marine plant which is obviously a large kelp, or which is olive-drab to military greenish-brown in color, is probably a brown alga (Phaeophyta). The red algae may be distinctly pink or red, or vary to deep purplish-brown or nearly black. Many members of the Rhodophyta are relatively small, others are large and conspicuous. Assuming you have drawn the correct conclusion as to the division, the following keys and drawings will help you to identify some of the more common algae of northern California shores.

Key to Some Common Green Algae of Northern California

1. Body of plant relatively smooth, without projecting hairlike filaments or felty surface 2
1. Body of plant either felty, "spongy", or composed of hair-like filaments... 3
2. Body of plant composed of thin, bright-green, broad sheets, often with ruffled edges; on rocks, pilings, floats, etc. .... *Ulva* spp. (fig. 156)
2. Body of plant not sheet-like, composed of elongate, soft, somewhat twisted tubes; found in rocky intertidal... *Enteromorpha* sp.
3. Plant forms clumps of soft spongy green tufts; without well-developed erect branches; found on mid- to low-intertidal rocks ... . *Cladophora* sp.
3. Plant not as above; some portion of plant forming elongate erect branches 4
4. Erect portion of plant composed of hair-like filaments; found in mid- to low-rocky intertidal............... *Spongomorpha coalita*
4. Erect portion of plant composed of fleshy, somewhat spongy and felt-like branches; found in rocky intertidal .... *Codium fragile* (fig. 157)

Key to Some Common Brown Algae of Northern California

1. Plants resembling small palm trees, about .5m high; restricted to wave-swept rocky shores........ *Postelsia palmaeformis* (fig. 158)
1. Plant not as above 2
2. Plants not exceeding .3m (about 1ft) in length; thalli display regular patterns of branching; the rockweeds (figs. 159-160) 3
2. Plants usually exceeding .3m in length; branching variable...... 4
3. Blades distinctly flattened, each with an obvious midrib; found in upper- mid-intertidal zones on rocky shores .... *Fucus distichus* (fig. 159)
3. Blades not extremely flattened, lacking an obvious midrib along each blade; found in upper-mid-intertidal zones on rocky shores... .... *Pelvetiopsis limitata* (fig. 160)
4. Large nearshore kelp; whip-like stipe (up to 3-4m long) ends in hollow bulb-like float from which numerous elongate blades arise .... *Nereocystis luetkeana* (fig. 161)
4. Not as above 5
5. Body of plant composed of regularly-branched but often fragmented blades; often several meters in length; small floats arise from ends of branches as strings of several adjacent spheres; found in mid- to low-rocky intertidal and less frequently in bays .. .. *Cystoseira osmundacea* (fig. 162)
5. Body not as above; floats never like strings of tiny bead-like spheres 6
6. Plant with distinct, elongate, strap-like stipe from which small blades and oval floats arise along each side; stipe up to several meters in length; common in lower-mid-intertidal zones of rocky shores......................... *Egregia menziesii* (fig. 163)

6. Body not as above 7
7. With one or more blades (often broken) arising from the end of a stiff short (up to .5m) stipe; common in surge channels of low rocky-intertidal zones .... *Laminaria* spp.
7. With short stipe which extends as a flattened midrib along the midline of a very elongate, sheetlike blade; several meters in length; found in lower-mid-intertidal of rocky shores, and on tidal flats; **extremely** slippery when exposed at low tides............. .... *Alaria marginata* (fig. 164)

Key to Some Common Red Algae of Northern California

1. Body of plant crusty (calcareous), either growing in short erect patches or encrusting flat on solid substrates; usually bright pink or, if dead, white; common throughout the mid- to low-rocky-intertidal zones... various coralline algae *(Corallina and Bossiella)*
1. Body not calcareous.. 2
2. Plants form tar-like, deep-red or brown to black encrustations on high- and mid-intertidal rocks .... *Hildenbrandia occidentalis* and *Petrocelis franciscana*
2. At least some portion of the plant erect as blades or filaments ... 3
3. Body of plant in the form of reddish-brown, rather stiff branching tufts a few centimeters high; common in the high tide zone of rocky shores.......................... *Endocladia muricata*
3. Body and habitat not as above 4
4. Body of plant composed of elongate, finger-like, water-filled bags, several centimeters in length .... *Halosaccion glandiforme*
4. Plant not as above 5
5. Plant branched or leaf-like; usually deep-red to nearly black in color; surface usually covered with bumps or "warts" (the so-called "turkish-towel" algae)...... *Gigartina* spp. (A great variety of body forms is displayed by members of this genus making certainty of generic or specific recognition difficult.)
5. Plant body composed either of filaments or of smooth blades without numerous bumps on the surface.......................... 6
6. Plant body delicate and composed of numerous tiny filaments; usually brick-red in color; less than 25cm in length; one species may frequently be found attached to the siphons of the horseneck clam, *Tresus* .... *Polysiphonia* spp.
6. Plants large or small, but not composed of filaments; body composed of smooth, flattened blades.......................... 7
7. Blades usually less than 4cm high; growing as red blades on the surface of *Zostera* (eelgrass) and *Phyllospadix* (surf grass)........ .... *Smithora naiadum*
7. Plants much larger than 4cm high, attached to hard substrates .. 8
8. Plants composed of flattened ruffled blades, usually 10 or more centimeters in length; greenish mixed with dull red color; common in upper-mid-rocky-intertidal zones .... *Porphyra* sp.
8. Blades larger (up to about .5m long), smooth, and oval; surface of blades iridescent, oily in appearance when submerged; found in lower tidal pools on rocky shores.................... *Iridaea* sp.

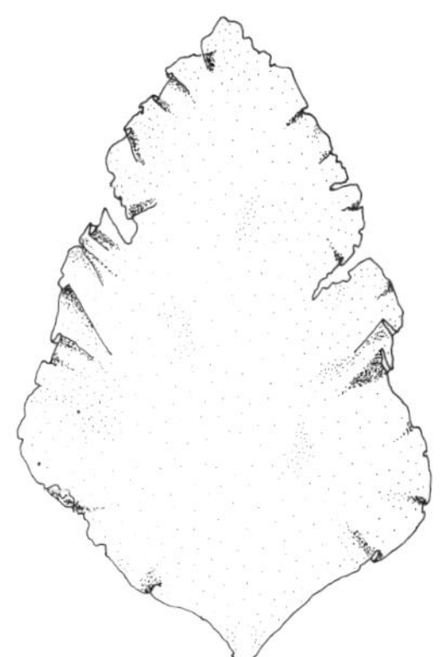

156. *Ulva* (up to 30cm long)

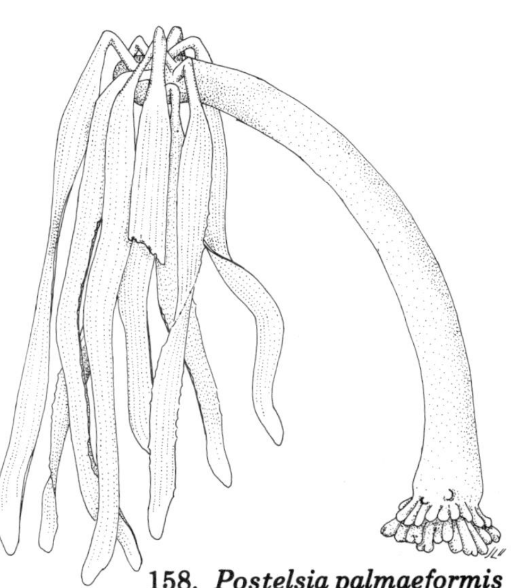

158. *Postelsia palmaeformis* (~25cm long)

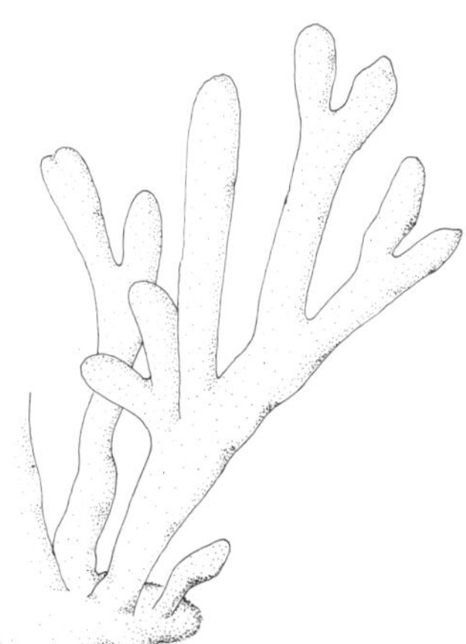

157. *Codium fragile* (~15cm long)

159. *Fucus distichus* (~15cm long)

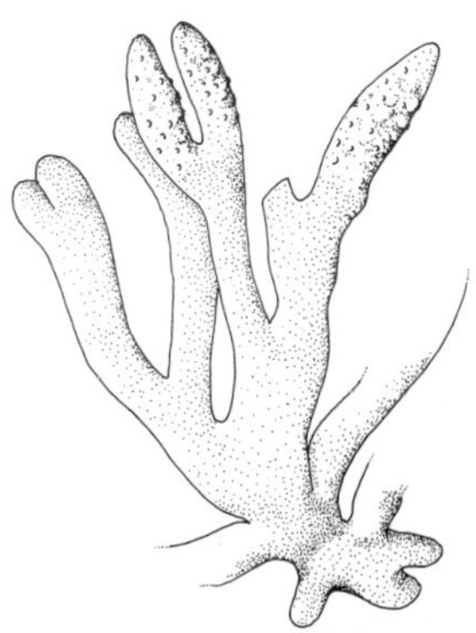

160. *Pelvetiopsis limitata* (~8cm long)

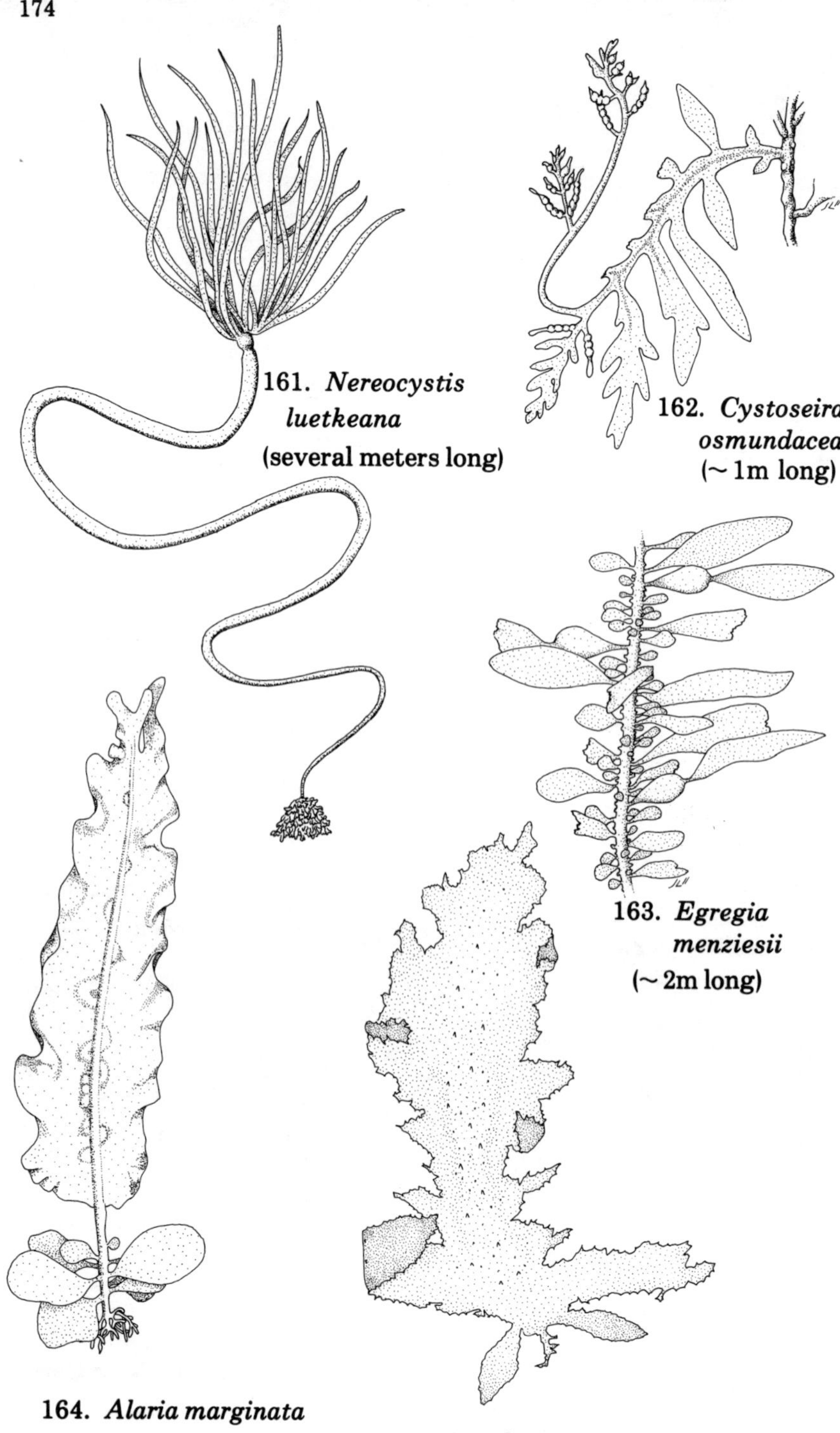

161. *Nereocystis luetkeana* (several meters long)

162. *Cystoseira osmundacea* (~ 1m long)

163. *Egregia menziesii* (~ 2m long)

164. *Alaria marginata* (~ 3m long)

165. *Gigartina* (up to 30cm long)

APPENDIX 1

Metric Conversions

Volume:

1 quart = 32 fluid ounces = 946 milliliters = .946 liters
1 liter = 1000 milliliters = 33.8 fluid ounces

Weight:

1 pound = 454 grams = .454 kilograms
1 kilogram = 1000 grams = 2.2 pounds

Length:

1 meter = 1000 millimeters = 100 centimeters = 39.37 inches = 3.281 feet
1 millimeter = .1 centimeter = .039 inches = .0032 feet
1 centimeter = 10 millimeters = .394 inches = .0328 feet
1 inch = 25.4 millimeters = 2.54 centimeters = .0833 feet

Temperature:

$$°\text{Fahrenheit} = 1.8\,(°\text{Centigrade}) + 32$$

$$°\text{Centigrade} = \frac{°\text{Fahrenheit} - 32}{1.8}$$

APPENDIX 2

Classification Scheme of Animals and Plants

The following is but one of several possible ranking systems devised by biologists.

Kingdom Animalia (the animals)
- Phylum Protozoa (the single-celled animals)
  - Subphylum Sarcomastigophora
    - Superclass Mastigophora (the flagellates)
    - Superclass Sarcodina (amebae, moving by means of cellular extensions called pseudopodia)
  - Subphyla Sporozoa and Cnidospora (odd protozoans which are parasitic in a variety of vertebrate and invertebrate hosts)
  - Subphylum Ciliophora (the ciliates)
- Phylum Porifera (the sponges)
  - Class Calcarea (the calcareous sponges)
  - Class Hexactinellida (the "glass" or siliceous sponges)
  - Class Demospongiae (sponges containing siliceous spicules and/or spongin fibers; includes the commercial bath sponges)
  - Class Sclerospongiae (strange coral-reef dwelling sponges with siliceous spicules, spongin, and an outer casing of calcium carbonate)
- Phylum Cnidaria (sea anemones, jellyfishes, corals, etc.)
  - Class Hydrozoa (the hydroids and hydromedusae)
  - Class Scyphozoa (the true jellyfishes or scyphomedusae)
  - Class Anthozoa (the sea anemones and corals)
- Phylum Ctenophora (the comb jellies and relatives)
- Phylum Mesozoa (minute parasites in the kidneys of cephalopods and body fluids of a variety of other invertebrates)
- Phylum Platyhelminthes (the flatworms)
  - Class Turbellaria (the free-living flatworms)
  - Class Cestoda (the tapeworms)
  - Class Trematoda (the flukes)
- Phylum Nemertea (the ribbon worms)
- Pseudocoelomates: seven phyla are included here containing such things as roundworms, rotifers, and a variety of other creatures.

Phylum Annelida (the segmented worms)
Class Polychaeta (the clam worms, tube worms, etc.; most of the marine annelids are included in this class)
Class Oligochaeta (the earthworms and their kin)
Class Hirudinea (the leeches)
Phylum Sipuncula (the peanut worms)
Phylum Echiura (the spoon worms, and *Urechis*)
Phylum Priapula (mud-dwelling worms once considered pseudocoelomates, but no longer)
Phylum Pogonophora (the beard worms; a group of predominantly deep-water, tube-building worms)
Phylum Arthropoda (the arthropods)
Subphylum Chelicerata
Class Merostomata (the east coast horseshoe crabs)
Class Arachnida (spiders and their relatives)
Class Pycnogonida (the "sea spiders")
Subphylum Mandibulata
Class Insecta (the insects)
"Class" Myriapoda (the centipedes, millipedes, etc.; often separated into several classes)
Class Crustacea
Subclass Cephalocarida (tiny blind creatures, living between sand grains on quiet beaches)
Subclass Branchiopoda (an odd assemblage of primitive, primarily fresh water crustaceans; the brine shrimps are included here)
Subclass Ostracoda (these animals are characterized by having their bodies enclosed in a bivalved carapace like clam shells; they are extremely abundant in the sea; the intertidal ones are very small)
Subclass Mystacocarida (minute, somewhat vermiform, crustaceans living between sand grains)
Subclass Branchiura (crustaceans parasitic on the skin of fishes)
Subclass Copepoda (small, extremely abundant crustaceans, of great importance in the sea's food chains)
Subclass Cirripedia (the barnacles)
Subclass Malacostraca (some minor groups are omitted from the listing below)
Series Phyllocarida (the nebaliids)
Series Eumalacostraca
Superorder Hoplocarida (the stomatopods)

Superorder Peracarida

Order Mysidacea (the opossum shrimps)

Order Isopoda (the rock lice, sea slaters, pill bugs, etc.)

Order Amphipoda (beach hoppers and their relatives)

Order Cumacea (strange little benthic peracarids; they are extremely abundant in local subtidal sands)

Order Tanaidacea (something like isopods, but with a short carapace and tiny claws)

Superorder Eucarida

Order Euphausiacea (krill)

Order Decapoda (crabs, lobsters, shrimps, etc.)

Phylum Tardigrada (the water bears; minute creatures common in damp moss and in fresh water)

Phylum Onychophora (terrestrial, caterpillar-like creatures closely related to arthropods)

Phylum Mollusca (the mollusks)

Class Monoplacophora (deep-water, somewhat limpet-like primitive (?) mollusks; the famous segmented "snail")

Class Aplacophora (shell-less, vermiform, benthic creatures)

Class Polyplacophora (the chitons or sea cradles)

Class Gastropoda (snails, sea slugs, limpets, etc.)

Class Bivalvia (clams, oysters, scallops, and mussels)

Class Scaphopoda (the tusk shells)

Class Cephalopoda (octopuses and squids)

Phylum Phoronida (the phoronid worms)

Phylum Ectoprocta (the bryozoans)

Phylum Brachiopoda (the lamp shells)

Phylum Chaetognatha (the planktonic arrow worms)

Phylum Echinodermata

Class Asteroidea (the sea stars)

Class Echinoidea (the sea urchins and sand dollars)

Class Ophiuroidea (the brittle or serpent stars)

Class Holothuroidea (the sea cucumbers)

Class Crinoidea (the sea lilies and feather stars)

Phylum Hemichordata (the tongue and acorn worms)

Phylum Chordata
- Subphylum Urochordata (the tunicates: salps and ascidians)
- Subphylum Cephalochordata (the lancelets)
- Subphylum Vertebrata (the vertebrates)
  - Class Agnatha (the lampreys and hagfishes)
  - Class Chondrichthyes (the sharks and rays)
  - Class Osteichthyes (the bony fishes: trout, salmon, perch, etc.)
  - Class Amphibia (frogs, toads, and salamanders)
  - Class Reptilia (snakes, lizards, and other creepy types)
  - Class Aves (the birds)
  - Class Mammalia (the mammals)

Kingdom Plantae (only those discussed in the text are included here)
- Division Pyrrophyta (the dinoflagellates)
- Division Chrysophyta (the diatoms)
- Division Chlorophyta (the green algae)
- Division Rhodophyta (the red algae)
- Division Phaeophyta (the brown algae, including kelps)
- Division Trachaeophyta (the vascular plants)

GLOSSARY

A:

abdomen: noun-phrase; the posterior group of segments of the body of arthropods.

aboral: adj.; referring to that region of a radially-symmetrical animal's body which is opposite from the region of the mouth.

acoelomate: adj.; without a coelom or other body cavity between the outer surface and the digestive tube (e.g. flatworms).

algin: noun; an extract of certain seaweeds, of commercial importance as an emulsifier.

ambulacral: adj.; referring to those areas on echinoderms which bear the tube feet, such as the ambulacral grooves on the undersides of the arms of seastars.

annulate: adj.; composed of or having the appearance of rings or ring-like structures, as the body of a segmented worm.

anoxic: adj.; without oxygen.

antennae: noun; anteriorly-placed sensory structures (usually sensitive to touch or "taste"); generally long and thin (as in crustaceans).

anterior: adj.; towards the front or head end of an animal's body.

aperture: noun; an opening.

arborescent: adj.; bushy, plant-like; like a branching tree.

asexual reproduction: noun-phrase; production of new individuals without sex, that is, without fertilization; as in reproduction by budding.

avicularium; noun; a modified zooid (individual) in bryozoan colonies, shaped like a bird's head and beak; functions to keep the colony surface clean and free of debris.

B:

baleen plates: noun-phrase; the comb-like or brush-like filtering structures in the mouths of baleen whales, used to extract food from the water.

benthic: adj.; referring to the sea bottom at any depth; organisms which reside there are collectively called the benthos.

bilateral symmetry: the form of symmetry in which an animal's body has but a single plane which separates it into halves which are mirror images of each other; that is, the animal has a right and left side and a front and back end.

biogeography: noun; the science dealing with the geographic distributions of organisms.

biomass: noun; a quantitative measurement (usually by weight) of the amount of living material in a given area.

biradial symmetry: the form of symmetry in which an animal's body has two planes which separate it into two mirror-image-like halves (as in ctenophores).

blades: noun; the "leafy" portions of an alga.

bloom: noun; the rapid growth and reproduction of plankton.

budding: noun; an asexual reproductive process whereby new individuals are produced from pre-existing individuals by vegetative growth rather than by sexual reproduction.

byssus: noun; the tough threads produced by certain bivalved mollusks (e.g. mussels) by which they attach to the substrate.

C:

calcareous: adj.; composed of calcium carbonate; generally hard and/or crusty, as clam shells or coralline algae.

carapace: noun: that portion of the exoskeleton of some crustaceans which is produced as a shield-like covering over the head and all or part of the thorax.

carnivore: noun; an organism which feeds upon living or dead animal matter.

central disc: noun-phrase; as in seastars; that portion of the body from which the arms radiate.

cephalic: adj.; pertaining to or associated with the head.

cephalon: noun; head.

cerata (singular ceras): noun; processes along the back of many sea slugs (nudibranchs).

chelae (singular chela): noun; pincer-like claws.

chelate: adj.; bearing chelae.

chelicerae (singular chelicera): noun; the first pair of appendages on the heads of such animals as arachnids and horseshoe crabs; generally pincer-like.

choanocytes: noun; the flagellated "collar cells" lining some or all of the internal spaces in sponges; function in creating the water currents through the sponge.

chromatophores: noun; pigment-containing cells.

cilia (singular cilium): noun; tiny hair-like projections from certain cells, capable of whip-like movement to create water currents, to move objects along cell surfaces, or to propel the animal (as in certain protozoans and tiny larvae).

cnidocytes: noun; cells in cnidarians which contain the "stinging threads" or nematocysts.

coelom: noun; a fluid-filled body cavity completely lined with a thin layer of tissue derived from mesoderm; cavity lies between the gut and the outer body wall.

coelomate: adj.; possessing a coelom.

colloblasts: noun; mucus-producing cells on the tentacles of ctenophores; used in prey capture.

commensalism: noun; a symbiotic relationship or association in which the symbiont (commensal) benefits and the host is unharmed (e.g. the various guests in the burrow of *Urechis*).

competition: noun; the utilization of some particular environmental commodity (food, space, etc.) by individuals of the same species (intraspecific) or members of two or more different species (interspecific), when that commodity is in short supply.

compound eye: noun-phrase; a type of eye unique to arthropods; composed of multiple light-sensitive units called ommatidia.

compressed: adj.; flattened from side to side (laterally), like amphipods and most fishes.

concentric: adj.; in a pattern consisting of rings of increasing diameter, one inside the other; such as the concentric growth rings on a clam's shells.

coralline algae: noun-phrase; a collective term for certain groups of red algae which are impregnated with calcium carbonate.

D:

dactylozooid: noun; a non-feeding member of certain hydroid colonies, armed with nematocysts and specialized for defense.

depauperate: adj.; impoverished, poorly developed.

depressed: adj.; flattened from top to bottom (dorsoventrally), like isopods.

dermal gills: noun-phrase; thin outpocketings of skin and coelomic lining in echinoderms, which function in gas exchange.

detritivore: noun; an animal which feeds upon detritus.

detritus: noun; loose, tiny bits and pieces of animal and plant bodies produced by mechanical breakdown or abrasion; provides an important source of food for many beasts in the sea.

dichotomous: adj.; a type of branching in which a branch divides into two equal or similar branches, or progressively divides in such a manner.

dioecious: adj.; having separate sexes, with male and female reproductive systems housed in separate organisms.

diploblastic: adj.; composed of two embryonic germ layers (ectoderm and entoderm).

diurnal tides: noun-phrase; type of tidal pattern in which there is one high tide and one low tide each day.

dorsal: adj.; referring to or related to the back or backside of an animal.

dorsum: noun; the back or backside of an animal.

E:

ecdysis: noun; the periodic shedding of the exoskeleton in arthropods; molting.

ecology: noun; the study of the relationships between organisms and their environment.

ectoderm: noun; the embryonic germ layer responsible for the formation of the outer skin and its derivatives, and the nervous system.

embryogeny: noun; the development of an individual from a fertilized egg to a juvenile organism.

embryology: noun; the science dealing with the study of the development of organisms from the fertilized egg to the juvenile stage.

encrusting: adj.; forming a thin crust-like layer on another object.

entoderm: noun; the embryonic germ layer responsible for the formation of the digestive tract and its associated structures.

epifauna: noun; a collective name for those animals living on the surface of the substrate; a subcategory of benthos.

errant: adj.; possessing the ability to crawl or walk about from one place to another; not attached.

eversible; adj.; capable of being protruded by turning inside out.

excretion: noun; the process of eliminating nitrogenous waste products from the body, usually in the form of ammonia in most marine animals.

exoskeleton: noun; a skeleton which forms the outermost covering of the body, as in arthropods.

F:

feces: noun; undigested food material, bacteria, etc., passed through the gut and out the anus.

fertilization: noun; the union of male and female gametes (egg and sperm) to produce a zygote.

filamentous: adj.; threadlike or consisting of threadlike parts.

filter feeder: an animal which obtains its food by filtering or straining the water by any of a variety of means.

flagella (singular flagellum): noun; long thread-like or hair-like extensions from cells, which are essentially very long cilia.

food chain/web: noun-phrase; the passage of energy through the various living components of an ecological system; production of organic compounds by plants, which are eaten by herbivores, which are in turn eaten by predators, etc.

G:

gametes: noun; male or female sex cells; sperm and eggs.

gametophyte generation: the stage in a plant's life cycle which produces gametes.

gastrozooid: noun; feeding individuals in hydroid colonies; hydranths.

generic name: noun-phrase; first name of a species' binomial; is followed by the trivial name.

germ layers: embryonic tissue layers (ectoderm, entoderm, and mesoderm) from which all adult structures are derived.

gill: noun; any body surface or special structure modified for gas exchange in aquatic animals.

girdle: noun; the fleshy or leathery portion of the mantle of a chiton which forms a border on the dorsum around the shell plates.

globose: adj.; spherical or globular.

gonozooid: noun; non-feeding individuals in hydroid colonies whose usual function is to produce medusae by budding.

H:

hemocoel: noun; body cavity in arthropods and mollusks which is not lined by mesoderm, and in which the organs are suspended and bathed by blood.

herbivore: noun; an animal which feeds on plants or plant material.

hermaphroditic: adj.; synonym of monoecious; condition in which a single individual has both male and female reproductive systems.

heteronomous: adj.; a condition in segmented animals (annelids and arthropods) in which sets of segments are different from one another, being regionally specialized for different functions (e.g. the head, thorax, and abdomen of shrimps).

holdfast: noun; the "root-like" attachment structures of algae.

homonomous:adj.; a condition in segmented animals in which all the main body segments are more or less alike (e.g. the trunk of an errant polychaete).

hydranth: see gastrozooid.

I:

infauna: noun; collective name for animals living within the substrate; a subcategory of benthos.

inorganic: adj.; referring to material which is not composed of organic (carbon-based) compounds (e.g. calcium carbonate).

intertidal region: noun-phrase; roughly the shoreline; specifically that region from the level of the highest to the level of the lowest tides; littoral.

interstitial: adj.; referring to the spaces between sediment particles, or to the animals which inhabit those spaces (e.g. interstitial worms).

introvert: noun; in sipunculans, the elongate, narrow portion of the body which bears the mouth, and which is capable of being inverted inside the body, in a manner similar to pressing in the finger of a glove.

invertebrate: noun; an animal which lacks a backbone or vertebral column.

K:

kelp: noun; any of the relatively large brown algae.

key: noun; an organized tabulation of diagnostic characters arranged in such a way so as to facilitate identification of specific groups of animals or plants.

krill: noun; planktonic shrimp-like crustaceans of the Order Euphausiacea; important as food for many fishes and certain whales.

L:

lateral: adj.; referring to the sides of an animal or plant.

littoral: see intertidal.

local population: noun-phrase; individuals of the same species in a specific locality; a potentially interbreeding group.

longitudinal: adj.; oriented along the long axis of an organism's body or any elongate structure; e.g. longitudinal stripes on a worm.

lophophore: noun; a band or ring of hollow, ciliated tentacles partially or wholly encircling the mouth but not the anus, and each tentacle containing an extension of the animal's coelom; characteristic of the "lophophorate" phyla, Phoronida, Ectoprocta, and Brachiopoda.

M:

macroscopic: adj.; generally, large enough to be seen and recognized without the aid of any magnification device.

madreporite: noun; in many echinoderms, the plate-like sieve usually on the aboral surface; opens to the water-vascular system.

mandibles: noun; chewing mouth parts characteristic of insects, myriapods, and crustaceans.

mantle: noun; a fold of the body in mollusks responsible for the secretion of the shell.

mantle cavity: noun-phrase; in mollusks; the space between the body proper and the mantle fold(s), which houses the gills and receives the gametes, feces, and excretory products.

marsupium: noun; a pouch for brooding embryos or young; as occurs in the peracarid crustaceans.

maxillae (singular maxilla): moun; mouth parts in crustaceans; two pairs occur immediately posterior to the mandibles; function in food handling or filtering.

maxillipeds: noun; mouth parts in crustaceans derived from modified anterior thoracic legs; function in feeding.

medial: adj; in bilaterally symmetrical animals, referring to the direction towards the plane which divides the body into right and left halves; towards the midplane.

medusae (singular medusa): noun; the bell- or dish-shaped, sexual, pelagic stages in the life cycles of most Hydrozoa and Scyphozoa; jellyfishes.

mesenchyme: noun; the middle layer, between skin and gut, in most cnidarians and sponges.

mesoderm: noun; the middle germ layer (between ectoderm and entoderm) in triploblastic animals; gives rise to some of the musculature, circulatory system, gamete forming tissue, the lining of the coelom (when present), etc.

mixed tides: noun-phrase; semidiurnal tides in which the two highs and two lows are unequal.

monoecious: adj.; see hermaphroditic.

mutualism: noun; a symbiotic relationship or association in which both the symbiont and the host benefit.

N:

neap tides: noun-phrase; those tides during the monthly cycle with the lowest highs and the highest lows; tides of low amplitude.

nematocysts: noun; "stinging" threads located within certain types of cells (cnidocytes) in cnidarians; function in prey capture, defense, and attachment.

niche: noun; the position of an organism in the ecological scheme of things; its role in the food chain, its environmental requirements, etc.; may be called the organisms's function in the balance of Nature.

nomenclature: noun; a system of names.

notochord: noun; in chordates; a longitudinal dorsally-located stiffening rod in the body, often present only in embryonic or larval stages; replaced by the backbone in land vertebrates.

O:

omnivore: noun; an animal which feeds upon both plant and animal matter.

oostegites: noun; flaps on the ventral side of the thorax of peracarid crustaceans which form the marsupium or brood pouch; they are outgrowths of the thoracic legs.

operculum: noun; a horny or calcareous plate for closing over an aperture, as in some gastropods.

oral: adj.; pertaining to the mouth or the side of an animal on which the mouth is located.

organic: adj.; referring to or composed of chemical compounds which contain carbon and which are generally associated with living systems (e.g. proteins, fats, carbohydrates); as opposed to inorganic.

oscula (singular osculum): noun; the openings in sponges through which water leaves the internal cavities.

ostia (singular ostium): noun; the openings in sponges through which water enters the internal cavities.

P:

parapodia (singular parapodium): noun; the unjointed, often flap-like, paired appendages placed along the sides of the segments of polychaete worms; often aid in locomotion

parasitism: noun; a symbiotic relationship or association in which the symbiont (parasite) benefits at the expense of the host (e.g. the parasitic isopods on the gills of fishes).

pedicellariae: noun; tiny jaw-like structures on the body surfaces of echinoderms which function primarily to keep the body surface free from settling material.

pedicle: noun; a stalk; as in brachiopods.

peduncle: noun; a stalk; usually fleshy or leathery; as in lepadomorph barnacles.

pelagic: adj.; referring to the environment of the water not associated directly with the bottom; or to designate the organisms which live therein (e.g. pelagic fishes).

pereopods: noun; the jointed walking, swimming, or grasping legs on the thorax of crustaceans.

periostracum: noun; an outer, organic, layer on the shells of some mollusks; usually brown or black, and thin and papery.

phylogeny: noun; the evolutionary history or lineage within a group of organisms.

plankton: noun; the collective term for animals and plants (usually small) which drift in the water and are too weak to swim against ocean currents; zooplankton and phytoplankton respectively.

planula: noun; the characteristic oval, ciliated larval stage of cnidarians.

pleopods: noun; the swimming or other abdominal appendages of crustaceans.

polymorphism: noun; the condition wherein the members of a species exhibit several body forms; e.g. the various zooids in a hydroid colony.

polyp: noun; the attached (usually) stage in cnidarian life cycles; e.g. hydroids, corals, sea anemones.

population: noun; the collective term for members of a single species occurring within defined geographic boundaries.

posterior: adj.; towards the tail end or back end of bilaterally symmetrical animals.

predation: noun; the act of killing and eating another animal.

predator: noun; an animal which kills and eats other animals.

prehensile: adj.: referring to any appendage which is modified for grasping.

prey: noun; the collective term for animals which are killed and eaten by another animal.

proboscis: noun; an elongate, flexible, anteriorly located snout; or part of the foregut capable of being everted; usually functions in feeding.

productivity: noun; generally, the rate of production of organic matter by photosynthesis.

protandry: noun; a condition in which an individual is first male and then female.

protrusible: adj; capable of being thrust out away from the body, as the extension of a tentacle.

pseudocoelom: noun; a body cavity between the body wall and the gut which is not fully lined by membranes derived from mesoderm.

pseudocoelomate: adj.; possessing a pseudocoelom.

R:

radial symmetry: noun-phrase; having the body parts arranged around the axis in such a way that the body may be cut along its axis on any plane to yield two halves which are mirror images of each other.

radiating: adj.; lines, markings, ridges, etc., which are arranged so as to extend away from a common point of origin.

radula: noun; the strap-like rasping feeding structure found in many mollusks.

respiratory tree: noun-phrase; the branching pouch off the hindgut of sea cucumbers which functions in gas exchange.

rhinophores: noun; the anteriorly placed sense organs in sea slugs (nudibranchs).

rostrum: noun; any snout-like extension of the head; particularly a forward extension of the carapace in some crustaceans.

rugose: adj.; folded or wrinkled.

S:

salinity: noun; generally, the amount of dissolved salts in water; usually expressed as parts per thousand (0/00) by weight; most ocean water averages about 34-35 °/00.

scavenger: noun; an animal which feeds primarily on dead animal and/or plant material.

scientific name: noun-phrase; the binomial designation of a species; includes the generic and trivial names.

scyphistoma: noun; the polyp stage of members of the Class Scyphozoa.

seaweeds: noun; macroscopic marine plants.

sedentary: adj.; remaining in one place much of the time, but not actually attached to the substrate.

segments: noun; the serially repeating sections of the bodies of annelids, arthropods, and vertebrates.

seiche: noun; the sloshing back and forth of water in partially or fully enclosed bays, lakes, or ocean basins.

semidiurnal tide (semidaily tide): noun-phrase; tidal pattern in which there are two highs and two lows each day.

sessile: adj.; permanently attached to the substrate; or, meaning non-stalked, as in balanomorph barnacles, and certain crustacean eyes.

setae (singular seta): noun; small thin spines, particularly those of polychaete parapodia, or the hair-like projections on arthropods.

siliceous: adj.; composed of silicon dioxide.

siphon: noun; a tube or opening which functions in carrying water in or out of the body or feeding chamber of an animal.

solitary: adj.; living singly; not in clusters or colonies.

species: noun; groups of actually or potentially interbreeding natural populations which are reproductively isolated from other such groups (Mayr, 1969, **Principles of Systematic Zoology**).

species abundance (species richness): noun-phrase; the number of kinds (species) of organisms in a particular area.

species diversity: noun-phrase; the combined values of species richness and species equitability; calculated using any of a number of mathematical formulae.

species equitability (evenness): noun-phrase; a measure of how evenly, in terms of numbers of individuals, the various species within a sample are distributed.

spring tides: noun-phrase; those tides during a monthly cycle with the highest highs and the lowest lows; tides of greatest amplitude.

specific epithet (trivial name): noun-phrase; the second term in a species' binomial; follows the generic name.

spicules: noun; the inorganic skeletal elements of certain sponges; may be siliceous (composed of SiO_2) or calcareous (composed of $CaCO_3$).

spongin: noun; the fibrous, organic, flexible material making up the skeleton of certain sponges.

stipe: noun; the "trunk-like" or "stem-like" portion of an alga; often strap- or whip-shaped.

strobila: noun; a scyphozoan polyp which is in the process of budding off medusae.

subtidal: adj.; referring to the area or the organisms below the level of the lowest tides.

supralittoral zone: noun-phrase; the area just above the level of the highest tides but which is directly affected by the rise and fall of the tides by receiving spray during the high tides.

symbiosis: noun; the living together of two different species in a constant and definite relationship; commensalism, mutualism, and parasitism.

systematics: noun; the study of the diversity and relationships of organisms; generally involves taxonomy.

T:

taxa (singular taxon): noun; groups of organisms designated by a name and rank within the classification scheme.

taxonomy: noun; the theory and practice of classifying organisms.

telson: noun; the terminal "plate" on the abdomen of most crustaceans; attached to the last segment of the body; often forms a tail-fan with the uropods.

tentacle: noun; any short or long, fleshy appendage on an animal; usually sensory, but may function in feeding or gas exchange; generally located at the anterior end of bilaterally symmetrical animals.

test: noun; particularly, the calcareous skeleton of echinoids.

thallus (plural thalli): noun; the main body of an alga, excluding the holdfast.

thorax: noun; the mid-body region of arthropods, especially crustaceans and insects, which bears the walking legs.

tidal flats: noun-phrase; gently sloping areas of mud or sand exposed at low tides, especially in areas of little or no wave action such as bays and estuaries.

torsion: noun; in gastropods; the twisting of the body mass 180° during the larval stage which brings the mantle cavity and associated structures to an anterodorsal position.

transverse: adj.; across, from side to side.

triploblastic: adj.; referring to the possession of three germ layers (ectoderm, entoderm, and mesoderm).

trivial name: see specific epithet.

trochophore: noun; the top-shaped ciliated larval type common in polychaetes, sipunculans, echiurans, and many mollusks.

tube feet: the hollow, sucker-like appendages connected to the water-vascular system in echinoderms which function in attachment, feeding, locomotion, etc.

tunic: noun; the outer body casing of urochordates.

U:

upwelling: noun; the movement of cold, nutrient-rich water from deeper layers towards the surface.

uropods: noun; the last pair of abdominal appendages in many crustaceans which often form a tail fan together with the telson.

V:

valves: noun; shells, or plates of shell material.

veliger: noun; the compact, shelled, swimming larva typical of mollusks, generally develops from a trochophore.

ventral: adj.; bottom or belly side; opposite of dorsal.

vermiform: adj.; long and relatively slender; wormlike.

vernacular name: noun-phrase; common name applied to an organism or group of organisms.

W:

water-vascular system: noun-phrase; in echinoderms; the water-filled (or coelomic fluid-filled) hydraulic system which operates the tube feet.

Z:

zooid: noun; an individual member of a colony.

zonation: noun; the occurrence of organized bands of organisms on the shore, with recognizable upper and lower borders; vertical stratification.

zygote: noun; a fertilized egg.

SELECTED REFERENCES

Abbott, R. 1968. A Guide to Field Identification: Seashells of North America. Golden Press, N. Y. 280pp. A good introduction to what mollusks and shells are all about.

Abbott, I. and G. Hollenberg. 1976. Marine Algae of California. Stanford Univ. Press, Stanford, Calif. 827pp. A long-awaited work, the most complete book on the California algae.

Allen, R. 1969. Common Intertidal Invertebrates of Southern California. Peek Publications, Palo Alto, Calif. 170pp. A book of keys to the critters of the southern end of the state.

Amos, W. 1958. Life in Pacific Tidepools. Doubleday, Garden City, N. Y. 45pp. A cut-and-paste book dealing mostly with the beaches around Monterey.

------- 1966. The Life of the Seashore: Our Living World of Nature. McGraw-Hill, N. Y. 231pp. Mostly Pacific stuff, a good introduction.

Arnold, A. 1901. The Sea-Beach at Ebb-Tide: A Guide to the Study of the Seaweeds and the Lower Animal Life Found Between the Tide-Marks. 490pp. Originally published by the Century Co., this classic is now available (1968) in one of the fine reprintings from the Dover Publications Co., N. Y. This book is excellent considering the time of its writing, but deals primarily with Atlantic species.

Bascom, W. 1964. Waves and Beaches: the Dynamics of the Ocean Surface. Doubleday and Co., Garden City, N. Y. 267pp. A solid, readable account of the subject; persists as the best introduction to the interactions of sea and shore.

Boyd, M. and J. DeMartini. 1977. The Intertidal and Subtidal Biota of Redwood National Park. Project Report, U. S. Department of the Interior, National Park Service. 162pp. The only comprehensive scientific account of the shores of the Redwood National Park in northern California; an excellent reference.

Braun, E. and V. Brown. 1966. Exploring Pacific Coast Tide Pools. California Naturegraph, Healdsburg, Calif. 56pp. There are better books for the beginning beachcomber, with much better pictures.

Brown, V. 1976. Sea Mammals and Reptiles of the Pacific Ocean. Collier Macmillan Ltd., London. 256pp. Descriptions, stories, and folklore.

Brusca, G., W. Mauck, and R. Meyer. 1971. The Stomatopod's Guide to the Common Seashore Life of Northern California. Special Publication No. 1, American Stomatopod Society, Eureka, Calif. 50pp. The several hundred copies are almost gone; apparently they found use in various quarters, and they continue to turn up in the oddest places.

Brusca, R. 1973. A Handbook to the Common Intertidal Invertebrates of the Gulf of California. Univ. of Arizona Press, Tucson, Ariz. 427pp. The only book of its kind dealing with the Sea of Cortez; it is presently being revised and enlarged.

Carefoot, T. 1977. Pacific Seashores, A Guide to Intertidal Ecology. Univ. of Washington Press, Seattle, Wash. 208pp. A new welcome addition to the literature on the Pacific Coast and the principles of seashore ecology in general. Carefoot has done a fine job of pulling a great deal of information together into a readable and informative account: a laudable performance!

Carl, G. 1963. Guide to the Marine Life of British Columbia. Handbook No. 21, British Columbia Provincial Museum, Victoria, B.C. 135pp. Mostly about vertebrates; the drawings are nice.

Cheng, L. 1976. Marine Insects. North-Holland Publ. Co., Amsterdam. 581pp. An entire book, well written and illustrated, devoted to the insects associated with the sea.

Dawson, E. 1956. How to Know the Seaweeds. Wm. C. Brown Co. Publishers, Dubuque, Iowa. 197pp. Keys and descriptions for coastal algae of both sides of the United States.

----- 1965. Marine Algae in the Vicinity of Humboldt State College. Available through the Humboldt State University Bookstore, Arcata, Calif. 76pp.

----- 1966. Seashore Plants of Southern California. Univ. of California Press, Berkeley and Los Angeles, Calif. 101pp.

----- 1966. Seashore Plants of Northern California. Univ. of California Press, Berkeley and Los Angeles, Calif. 103pp.

DeLuca, C. and D. DeLuca. 1976. Pacific Marine Life. A Survey of Pacific Ocean Invertebrates. C.E. Tuttle Co., Inc., Rutland, Vt. 66pp. A rather large undertaking to be crammed within sixty-six pages.

Daugherty, A. 1965. Marine Mammals of California. California Department of Fish and Game, Sacramento, Calif. 87pp. Good introductory information on seals, sea lions, whales, and the like.

Farmer, W. 1968. Tidepool Animals from the Gulf of California. Wesword Co., San Diego, Calif. 70pp. A useful handbook for the beginner in the Sea of Cortez; the drawings of the opisthobranchs are excellent.

Fitch, J. and R. Lavenberg. 1975. Tidepool and Nearshore Fishes of California. Univ. of California Natural History Guides, Berkeley and Los Angeles, Calif. 156pp. One of their continuing series; we await the book on sharks and rays. These are excellent books, but we wish the publishers had not reduced the size of the color photographs quite so much.

Flora, C. and E. Fairbanks. 1966. The Sound and the Sea: A Guide to Northwestern Neritic Invertebrate Zoology. Pioneer Printing, Bellingham, Wash. 455pp. Mostly around Puget Sound.

Gotshall, D. 1977. Fishwatcher's Guide to the Inshore Fishes of the Pacific Coast. Sea Challengers, Aqua-Craft, San Diego, Calif. 108pp. Good underwater photographs and accurate text by an ex-northern-California naturalist.

Guberlet, M. 1956. Seaweeds at Ebb Tide. Univ. of Washington Press, Seattle, Wash. 182pp. Many useful drawings with interesting natural-history notes; some of the vernacular names may lead to confusion.

Hancock, D. and L. Hancock. 1970. Wild Islands, Wildlife Adventures on North America's Pacific Coast. Wildlife Conservation Center, British Columbia. 79pp. Mostly about offshore birds and mammals; full of excellent photographs.

Hedgpeth, J. 1961. Common Seashore Life of Southern California. California Naturegraphs, Healdsburg, California. 65pp. Useful south of Point Conception.

----- 1962. Introduction to Seashore Life of the San Francisco Bay Region and the Coast of Northern California. Univ. of California Press, Berkeley and Los Angeles, Natural History Guides 9. 136pp. An excellent book to introduce you to the animals of our shores.

Hollenberg, G. and I. Abbott. 1966. Supplement to Smith's Algae of the Monterey Peninsula. Stanford Univ. Press, Stanford, Calif. 130pp.

Hewlett, S. and K. Hewlett. 1966. Sea Life of the Pacific Northwest. McGraw-Hill Ryerson Ltd., Toronto and Montreal. 176pp. A lovely table-top book of text and beautiful photographs.

Johnson, M. and H. Snook. 1927. Seashore Animals of the Pacific Coast. A real classic now available as a Dover, N.Y., reprint edition. 659pp. Many of the names have changed since the original publication of this work in 1927 by Macmillan Co., but the book still stands as one of the finest of its type.

Kozloff, E. 1973. Seashore Life of Puget Sound, the Strait of Georgia, and the San Juan Archipelago. J. J. Douglas, Ltd. Vancouver, B.C. 282pp, plus 28pp of color plates. An excellent book from one of the west coast's top biologists.

----- 1974. Keys to the Marine Invertebrates of Puget Sound, the San Juan Archipelago, and Adjacent Regions. Univ. of Washington Press, Seattle, Wash. 236pp. Includes many of the organisms found along northern California beaches.

MacGinitie, G. and N. MacGinitie. 1949. Natural History of Marine Animals. McGraw-Hill, N. Y. 473pp. A truly excellent source of information on how intertidal animals live.

Miller, D. and R. Lea. 1972. Guide to the Coastal Marine Fishes of California. Department of Fish and Game Bulletin 157, California Department of Fish and Game, Sacramento, Calif. 235pp. Finally reprinted and available again, with a pretty green cover. One of the very best works for identification of California's marine fishes.

Morris, P. 1966. A Field Guide to Shells of the Pacific Coast and Hawaii. Houghton Mifflin Co., Boston, Mass. 297pp. One of the best of the Petersen Field Guide Series; with many excellent black and white photographs.

Petersen, R. 1961. A Field Guide to Western Birds. Houghton Mifflin Co., Boston, Mass. 309pp. Includes an excellent section on shore birds of our coast.

Pike, G. 1956. Guide to the Whales, Porpoises and Dolphins of the North-East Pacific and Arctic Waters of Canada and Alaska. Fisheries Research Board of Canada, Circular 32, Nanaimo, B. C. 15pp. Includes some good drawings which will help in identifying these animals, even when you only see their backs from on board ship.

Ricketts, E., J. Calvin, and J. Hedgpeth. 1968. Between Pacific Tides. Fourth Edition. Stanford Univ. Press, Stanford, Calif. 614pp. There is no finer book available on the natural history of Pacific coast marine life and generally what goes on along our shores; a must for all beachcombers and students of marine biology. One can spend hours just browsing through the annotated literature lists.

Scammon, C. 1874. The Marine Mammals of the Northwestern Coast of North America. Another great book reprinted by Dover, N. Y., in 1968, with 319pp. A classic and scientific landmark of its time.

Smith, G. 1944. Marine Algae of the Monterey Peninsula. Stanford Univ. Press, Stanford, Calif. 622pp. The standard text on the subject until the 1976 presentation by Abbott and Hollenberg.

Smith, L. 1962. Common Seashore Life of the Pacific Northwest. California Naturegraphs, Healdsburg, Calif. 66pp. Mostly north of here.

Smith, R. and J. Carlton (eds.). 1975. Light's Manual: Intertidal Invertebrates of the Central California Coast. Third Edition. University of Calif. Press, Berkeley and Los Angeles, Calif. 761pp. The product of many experts and the persistence and hard work of the editors. Light's manual is essentially a set of keys written for those with some background in zoology; the best available book of its type for California and a must for all serious students who frequent the shore.

Tierney, R., J. Ulmer, L. Waxdeck, H. Foster, and J. Eckenroad. 1966. Exploring Tidal Life Along the Pacific Coast. Tidepool Associates, Berkeley, Calif. 64pp. Put together by a group of high-school biology teachers who frequented the Dillon Beach area for a number of summers. Fortunately, several of the authors are better at biology than they are at darts.

Vessel, M. and H. Wong. 1965. Seashore Life on Our Pacific Coast. Fearon, Palo Alto, Calif. 60pp. Does not compare favorably with many other similar publications.

Waaland, R. 1977. Common Seaweeds of the Pacific Coast. Pacific Search Press, Seattle, Wash. 120pp. A new and useful guide.

Yonge, C. 1949. The Seashore. Originally with 311pp. by Collins of London, has been reprinted (1963) in paper by Atheneum, N. Y., with 350pp. but lacking the lovely color plates of the original. All in all, one of the best accounts of how things are along the shore.

INDEX